Canine Pioneer

Canine Pioneer

The Extraordinary Life of
RUDOLPHINA MENZEL

Edited by Susan Martha Kahn

Brandeis University Press
WALTHAM, MASSACHUSETTS

Brandeis University Press
© 2022 Susan Martha Kahn
All rights reserved
Manufactured in the United States of America
Composed in Walbaum, Filosofia, and Frauen

For permission to reproduce any of the material
in this book, contact Brandeis University Press,
415 South Street, Waltham MA 02453,
or visit brandeisuniversitypress.com

LIBRARY OF CONGRESS CATALOGING-IN-PUBLICATION DATA
Names: Kahn, Susan Martha, 1963– editor.
Title: Canine pioneer : the extraordinary life of
Rudolphina Menzel / Susan Martha Kahn.
Description: Waltham, Massachusetts : Brandeis University Press, 2022. | Series: The Tauber Institute series for the study of European jewry | Includes bibliographical references and index. | Summary: "This book brings to light the story of Rudolphina Menzel, a fervent Zionist who was responsible for inventing the canine infrastructure in what came to be the State of Israel and for training hundreds of dogs to protect Jewish lives and property in pre-state Palestine" — Provided by publisher.
Identifiers: LCCN 2022025345 | ISBN 9781684581221 (paperback) | ISBN 9781684581214 (cloth) | ISBN 9781684581238 (ebook)
Subjects: LCSH: Menzel, Rudolphina, 1891–1973. |
Animal specialists—Israel—Biography. |
Psychologists—Israel—Biography. |
Zionists—Biography. |
Dogs—Israel—History. |
BISAC: BIOGRAPHY & AUTOBIOGRAPHY / Women |
HISTORY / Middle East / Israel & Palestine |
LCGFT: Biographies.
Classification: LCC QL31.M524 C36 2022 |
DDC 590.92 [B]—dc23/eng/20220713
LC record available at https://lccn.loc.gov/2022025345

5 4 3 2 1

This publication has been made possible
through the generous support of
Steven L. Rachmuth, Brandeis Class of 2014
and a lifelong dog lover.

*In memory of Carl, Lotty, David, and Otto Rachmuth,
and Gilda and Mayer Milstoc.*

To my daughter Esther, with great love

Contents

Preface   xi

Part I · Rudolphina Menzel's Legacy

1. The Extraordinary Life of Rudolphina Menzel   3
   SUSAN MARTHA KAHN

Part II · Perspectives on Rudolphina Menzel's Legacy

2. Rudolphina's Early Years in Austria   85
   MONIKA BAÁR

3. Rudolphina Menzel's Invention of
   Modern Dog Culture in Israel   91
   RACHEL KORIAT

4. Canine Zionism: Rudolphina Menzel
   and Working Dogs in Mandate Palestine   109
   BINYAMIN BLUM

5. Rudolphina Menzel's Contributions
   to the British War Effort   119
   LEA LEHAVI

6. Personal Recollections of Rudolphina Menzel
   and Her Canaan Dog Breed   123
   MYRNA SHIBOLETH

7. Rudolphina Menzel in Israeli Culture
   and Historiography   131
   TAMMY BAR-JOSEPH

Acknowledgments	137
Notes	139
Select Bibliography of Rudolphina Menzel's Publications	163
Contributors	169
Index	173

Photographs follow page 82.

Preface

RUDOLPHINA MENZEL (née Waltuch, 1891–1973) was a Viennese-born Jewish scientist whose pioneering research on canine psychology, development, and behavior fundamentally shaped the ways dogs came to be trained, cared for, and understood. Between the two world wars, Rudolphina was known all over Europe as one of the foremost researchers on canine cognition as well as among the most famous breeders and trainers of police dogs. Throughout the 1920s and until the Nazis seized power in 1933, she was a sought-after consultant at Kummersdorf, the German military dog training institute near Berlin. She was also a fervent Zionist who was primarily responsible for inventing the entire canine infrastructure in what came to be the State of Israel as well as for training hundreds of dogs to protect Jewish lives and property in pre-state Palestine. Teaching Jews to like dogs and training dogs to serve Jews became Rudolphina's distinctive Zionist mission.

In 1938 Rudolphina escaped Nazi-occupied Austria and moved to Palestine, where hundreds of dogs trained according to her methods served alongside Jewish forces in the 1948 war.[1] In the 1950s, Rudolphina created the first guide-dog institute in the Middle East and invented Israel's national dog breed, the Canaan dog. In 1962 at the age of seventy-one, she was appointed associate professor of animal psychology at Tel Aviv University, where she maintained an active research agenda almost until the day she died.[2]

Rudolphina's charisma, intelligence, and fortitude enabled her to create and sustain relationships with all manner of political, social, and professional colleagues around the world. As a young woman, she personally knew all the leading luminaries in the Austrian Social Democratic movement and was actively groomed for a leadership position in the party. As a scholar, she corresponded with many of the most well-known scientists of her day, including Albert Einstein and Ivan Pavlov. As a Zionist, she worked with a wide array of Zionist leaders, from Arthur Ruppin to Moshe Dayan. She witnessed Hitler's arrival in Austria as part of the Nazi annexation in 1938 and welcomed Helen Keller to Israel in 1952.

Any mention of Rudolphina Menzel must foreground the centrality of her relationship with her husband Rudolph, a successful physician in his own right. After meeting in a Zionist youth group at the University of Vienna, they lived and worked side by side for over fifty-five years. His unwavering support and devotion were critical to her professional success and personal happiness. What she loved, he loved. What she needed, he provided. Whatever aggravated her was ameliorated by him. He admired her greatly and was fiercely protective of his beloved "Dolphi." Theirs was a rare and deeply complementary partnership: he was quiet, warm, and methodical; she was irrepressible, charismatic, and disorganized. Their complementarity extended to the physical as well: he was tall, slim, and lanky; she was barely five feet tall and grew progressively rounder as she aged. The almost comical similarity of their names only added to the providence of their partnership. Though childless, they raised hundreds of dogs together.

Rudolphina published all of her work under both her and Rudolph's names in recognition not only of his steadfast financial support of her research but also of their unique style of collaboration. Rudolphina formulated theoretical questions, conceptualized research agendas, developed scientific methods, and generated all manner of data, while Rudolph imposed order on her results and ensured her conclusions were communicated clearly.[3] Both became masters of the burgeoning theoretical literature in cynology—the scientific study of dogs, which is a field that emerged in central Europe in the mid-nineteenth century. Cynologists were concerned with all topics related to *Canis familiaris*, from the origins of domestic dogs and canine evolution to the differentiation of dog breeds and canine cognition; many cynologists also bred dogs and were deeply concerned with practical questions regarding dog training techniques and methods. The Menzels became Austrian cynologists of international renown and published dozens of books and articles for both academic and popular audiences.

The Menzels' first and last books were for written for children. The first, entitled *Schwalbensommer* (Swallow Summer), was published in German in 1930. The last, entitled *Al kelavim, hatulim u-she'ar yedidim* (Dogs, Cats, and Other Friends), was published in Hebrew in 1968. Both are lively and accessible stories featuring familiar animals, written to instruct children how to better understand and relate to the animal world by describing the inner lives of birds, cats, and dogs and their capacities to feel human-like emotions. These didactic tales are fitting bookends for the Menzels' professional lives: at heart, they were educators for whom understanding the natural world was paramount to being human.

Rudolphina's papers include hundreds of assorted documents in Hebrew, German, and English in the Central Zionist Archives, the Archives of the Haganah (the Zionist military organization in pre-state Palestine), the National Library of Israel, the British National Archives, and the Leo Baeck Institute. She also published a range of scholarly works and left behind an unpublished German-language memoir, which she wrote between August 1939 and April 1940 as an entry to an essay contest sponsored by three Harvard professors (the contest was designed to solicit personal impressions from recent refugees about their life in Europe before and after Hitler assumed power in 1933; out of 230 entries, Rudolphina's received one of two second place prizes).[4] Menzel's contribution is summarized by the Harvard editors thus: "This extensive account by a woman with an unusual career describes her youth in a well-to-do Viennese family and the gradual growth of her Zionist consciousness. Her observations on dogs, scientists, Social Democracy, and other topics reveal her to be a believer in '*volkisch*' ideas about race and an upper-class eccentric. She emphasizes the personal friendship with many Germans she knew throughout the 1930s."[5] From these materials, combined with additional primary and secondary sources, as well as contemporary remembrances from people who knew the Menzels personally, we have assembled this volume.

In Part I, I explore how Rudolphina came to combine Zionism and dog training—an unusual amalgamation of political ideology, scientific expertise, and practical innovation that was uniquely her own. Here I outline the events, relationships, and experiences that converged to shape Rudolphina's worldview and priorities as she came of age as a young woman in Vienna. The subsequent section features an overview of Rudolphina's cynological career in Austria and her initial efforts to build a canine infrastructure in Palestine (1915–1938). The next section (1938–1948) begins with the Menzels' immigration from Austria to Palestine, when they were well into middle age, and details Rudolphina's intense involvement with Zionist institution-building in the decade leading up to Israeli statehood. The final section (1949–1973) offers a synopsis of Rudolphina's professional activities in the State of Israel.

In Part II, I present a series of perspectives on Rudolphina's legacy. Monika Baár provides a summary of Rudolphina's early years and places them in historical context. Rachel Koriat gives a detailed account of the canine infrastructure Rudolphina built in Palestine between 1938 and 1948. From Binyamin Blum, we learn about the relationship Rudolphina cultivated with the Palestine police as a consultant to their canine units during the British Mandate. Lea Lehavi explains the context in which Rudolphina contributed four hundred mine-detecting dogs to the British army for their use on the North African front during World War II. Myrna Shiboleth records her personal recollections of Rudolphina and describes her early efforts to establish the Canaan dog breed. Finally, Tammy Bar-Joseph elucidates a range of reasons Rudolphina has been overlooked as a key figure in Zionist state-building and has remained a footnote in Israeli historiography. Interestingly, Rudolphina's legacy has been more actively acknowledged in Europe: two scholarly articles on her life and work were recently published in German, and a street was named after her in Linz in 2016.[6]

Telling Rudolphina's story not only redresses the curious neglect of her scientific contributions and historical significance, it also brings into focus networks and connections that might otherwise have remained invisible. For example, we learn how service dogs trained by her and according to her methods conferred myriad advantages to Zionist settlement activities, both by protecting Jewish property and by saving Jewish lives. We see how Rudolphina's deep knowledge of dogs enabled her to solidify and deepen patronage relationships with the British Mandate government in ways that provided both tangible and intangible benefits to the Jews of Palestine. Her story also makes vivid different Jewish prejudices and predispositions towards dogs and pet-keeping more broadly, revealing new nuances of the intracultural tensions between Jews of different backgrounds, classes, and dispositions. The remarkable way Rudolphina rationalized the troubling overlaps between Nazi eugenics and the eugenic practices intrinsic to dog breeding raises added questions. Her invention of a national dog breed for Israel out of the indigenous dogs of Palestine reveals an additional wrinkle in the dynamics of colonialism. Further complicating her legacy is the fact that dogs trained according to her methods were routinely used as weapons against Jews in Europe and Arabs in Palestine—and yet, she subsequently trained dogs that provided vital services for both.

Preface

We hope this volume serves to illuminate important aspects of Rudolphina Menzel's life and career for an English-language audience and generates further scholarly interest in the scientific, political, and cultural contributions of this singular woman.

Susan Martha Kahn

PART I
Rudolphina Menzel's Legacy

ONE

The Extraordinary Life of Rudolphina Menzel

SUSAN MARTHA KAHN

Little in Rudolphina's background preordained either her Zionist awakening or her lifelong fascination with dogs. Born in 1891, she came of age in fin de siècle Vienna, that fabled era when Jewish acculturation produced a renowned explosion of creativity in the arts, sciences, literature, and music. This was the same cultural milieu that produced such assimilated Jewish luminaries as Sigmund Freud, Ludwig Wittgenstein, and Gustav Mahler; it was a fortuitous time and place for a precocious Jewish girl to be born into an affluent, cosmopolitan home.

Rudolphina depicted her earliest years as joyful and peaceful ones. Her family lived in a luxurious Viennese apartment that she described as a "comfortable and carefree island, where you could live far away from the battles of the world."[1] The youngest of four, Rudolphina was looked after by a personal nanny who indulged her every whim. Her father was a wealthy stockbroker and her mother, a beautiful, bourgeois housewife. In the winter, Rudolphina's parents got dressed up almost every night to go to parties, concerts, or the opera. In the summer, the happy family vacationed together in the Austrian countryside, swimming, socializing, and playing croquet.

This idyllic childhood evidently came to an abrupt end with a series of traumatic events that began in 1895 when Rudolphina was just four years old. First her mother died suddenly; then her father quickly remarried; and after that, her beloved nanny was fired. When her father lost a substantial sum in a bad business deal shortly thereafter, Rudolphina's entire world was turned upside down. Although the family managed to remain in their spacious Viennese apartment after these cataclysmic events, and Rudolphina's father eventually succeeded in rebuilding his fortune, there were apparently no more blissful, carefree summers in the countryside for the Waltuchs. Rudolphina reported these incidents as a matter of fact rather than a source of enmity. She reserved her resentment for her stepmother, whom she painted as a very unsympathetic woman who regarded her new stepchildren as nothing more than spoiled, bourgeois brats.

In her memoir, Rudolphina described her early exposure to Zionism as the result of her chance discovery of a discarded copy of the Zionist newspaper *Die Welt*, edited by Theodor Herzl—later known as the father of modern Zionism and a Viennese Jew himself. She explained how she "devoured" the publication and became enthralled by this new political movement and its foundational promise to restore the Jewish people to their natural and honorable condition by returning them to their ancient homeland in Palestine. That the movement was founded by assimilated Jews who "looked like her" rather than by traditional, religious Jews seemed to be particularly meaningful to her.

For an intelligent girl who had experienced such a profound rupture in the safety, security, and happiness she had known in her own early years, it is perhaps not surprising that Rudolphina found Zionist ideology deeply inspiring. The fundamental Zionist premise of working to normalize the Jewish condition and restore an idealized past may have resonated on more levels than she was aware of. Regardless of the psychological needs it may or may not have initially satisfied, Rudolphina's devotion to Zionism evolved from this early and accidental exposure to *Die Welt* into an enduring set of life-altering commitments.

Finding like-minded young Jews who shared her Zionist passions apparently wasn't easy. She decided against seeking out comrades at the local synagogue because she was resolutely anti-religious. She resisted joining a Jewish Sports Club since such associations typically catered to less affluent and cultured Jews than bourgeois Rudolphina. But she was desperate to find compatriots, so she put her elitism aside and joined a Jewish Gymnastics Club.

It seems that it was not a good fit. Most of the girls in the club just wanted to talk about clothes and love affairs in a middle-class jargon that Rudolphina found terribly coarse and offensive. She was shocked to discover that few knew anything about Zionism, and those who did mocked her for commitment to it; even her non-Jewish friends didn't ridicule her Zionism the way the "lower-class" Jews in the sports club did. (Indeed, Zionism was largely unpopular amongst Viennese Jews at the time—the idea of voluntarily exchanging theaters and concert halls for deserts and malarial swamps in order to revive some abstract notion of Jewish peoplehood seemed to many both unappealing and farfetched).[2] Rather than being put off by the other girls' initial resistance, charismatic Rudolphina took it upon herself to convert her fellow gymnasts to the cause. She created a "Young Judah" division in the club in order to infuse their exercise with the Zionist ethos of building strong, muscular Jews. She noted that the two girls who started out as her worst tormentors later became her closest friends; one even ended up immigrating to Palestine.

Not long after her accidental childhood conversion to Zionism, Rudolphina was introduced to socialism through another twist of fate she described as similarly fortuitous and transformative. Apparently, she found a brochure on her father's desk for a new school in Vienna called the Freie Schule. Founded in 1905 by bourgeois social reformers in the Socialist Democratic Workers Party, the school was designed as a radical new experiment meant to liberalize Austrian society by freeing it from the oppression of religious pedagogy.[3] She breathlessly described how exhilarating she found the description of the school's curriculum:

> The intellectual menu which was being offered here contained things that made me tremble with excitement. Among them were natural history and biology, hygiene and cultural history... For a few hours, I walked around like a dreamer, my mind was made up, *this* was the school I was going to attend, come what may![4]

At Rudolphina's insistence, her father sent her to the Freie Schule as soon as it opened. It seemed she loved just about everything about the school: how students were encouraged to think for themselves, voice their opinions, and explore topics like natural history and biology creatively and in-depth. The one course she dreaded was the required one in home economics. Already something of a snob and self-fashioned intellectual, Rudolphina expected the "handicraft" curriculum to be boring and the teacher to be old-fashioned. When a beautiful, smart young woman appeared at the head of the classroom, Rudolphina was not just shocked, she was smitten: "I think I was in love with her from the first moment," she noted.[5]

Frau Leopoldine Glockel turned home economics into a socialist object lesson about the importance of manual labor. She used instruction in sewing and knitting not only to impart practical skills but also as a pretext for getting to know the hearts and minds of her students. She was the first person who asked Rudolphina what she had to say, listened, and took her ideas seriously. When Rudolphina took it upon herself one day to stride around the room, lecturing the entire class on the imperatives of modern Zionism, Frau Glockel publicly thanked her for taking the time to explain something the class knew nothing about, instead of squelching the voluble passion of her opinionated, teenage student.[6]

Later that semester, Frau Glockel invited Rudolphina over for Christmas Eve dinner (the kind of personal invitation that is unforgettably thrilling to a smart young woman enthralled by an older female role model). The mere thought of spending the whole evening in the company of Frau Glockel, let alone being singled out from the rest of the class to do so, was almost too much for Rudolphina to bear. She noted: "The effect of this invitation on me was tremendous; on the one hand I was flushed with joy at the thought of being able to be with this person for a whole evening, on the other hand my old defiance awoke and it seemed impossible to me to go to a Christmas party... because I didn't want to celebrate someone else's festival."[7] So Rudolphina defiantly refused, explaining that first, she wasn't religious; second, she was Jewish; and third, she was not an assimilationist and would never celebrate a non-Jewish holiday. Again, rather than being put off by Rudolphina's strident rejection and impassioned explanation, Frau Glockel responded in a thoughtful way that left an opening for her young student; she patiently explained that she wasn't religious either and suggested instead that Rudolphina simply stop by for coffee on Christmas day. This was a compromise between her convictions and her desires that Rudolphina could tolerate, so off she went to Frau Glockel's house in the working-class Viennese district of Miedling.[8]

It turned out that Leopoldine was not only a home economics teacher. She and her husband Otto were well-known Viennese socialists and leaders in the Austrian Social Democratic Party, a party founded on an ideology of progressive, anticlerical socialism that played an influential role in fin de siècle Austria and throughout the early decades of the twentieth century.[9] Leopoldine was also an active feminist and published regularly in the radical party newspaper *Arbeiter-Zeitung*; Otto was an early proponent of women's rights and secular school reform who later served as the minister of education during the first Austrian Republic.[10] The Glockels quickly became like second parents to Rudolphina, and their lively, friendly home—which Rudolphina remembers as being always full of intellectual conversation—became both a refuge and an inspiration for her: a welcome alternative to the constant fights with her more limited and less worldly stepmother. That Leopoldine had also lost her mother when she was still a young girl may have created a particularly deep—and eventually lifelong—bond between the two women. Rudolphina later wrote that "everything that later became good and decent within me I ultimately owe to the luck of having found the motherly friendship of such a woman."[11]

The Glockels' salon was frequented by all manner of prominent activists in the Social Democratic Party; it was there that Rudolphina met Karl Seitz (1869–1950), who later became the first president of the Austrian Republic and longtime mayor of Vienna. She and Seitz became entangled in fraught debates about Zionism, with teenage Rudolphina outlining what she understood as its simple logic—that if the Austrians were a people with a national destiny, so too were the Jews—and thirty-six-year-old Seitz deriding all forms of nationalism as narrow-minded and chauvinistic. This was a time in Austrian history when belonging to a people, or *volk*, was seen as being at odds with the more universal promises of secular liberalism. Rudolphina remembered that Leopoldine often goaded her by asking why she couldn't be more like Viktor Adler, the Jewish leader of the Austrian Social Democrats, who had the good sense to reject all forms of Jewish nationalism. Rudolphina chose to respond by quoting a line from the Viennese Jewish poet Richard Beer-Hoffmann's famous poem "Lullaby for Miriam": "*Ufer nur sind wir, und tief in uns rinnt / Blut von Gewesenen.*" (We are but the banks of a river and the blood of our past flows wildly through us.) Her point was that the weight of Jewish history placed a special burden on Jews, which made it incumbent upon them to serve as channels for the Jewish past in order to sustain the Jewish future. Rudolphina seemed to take particular pleasure in recounting how she announced to the assembled leaders of the Austrian Social Democratic Party that this was a responsibility she was never going to abandon. She saw how Europe was organizing itself into distinct new nations as the old empires collapsed in the early years of the twentieth century, and she recognized the ways Jews were being excluded from these new forms of national belonging. To her, it was clear that Jews needed their own nation.

But managing her loyalties to Zionism, with its emphasis on particularistic national liberation, and her burgeoning sympathies with the universalism of Austrian social democracy presented a complicated, albeit imperative, intellectual task for adolescent Rudolphina. This task was made more difficult when Leopoldine and Otto, recognizing Rudolphina's brilliance, began grooming her for a political career as a future leader in the Social Democratic Party. Rudolphina remembered that Leopoldine implored her to "save her soul" by giving up Zionism and antiquated religious notions of Jewish particularity and to commit herself to building a more irreligious and liberal Austria as a Social Democrat. But despite the Glockels' repeated entreaties to run for a leadership position in the party, Rudolphina resisted; she felt her deeper allegiances precluded her from holding political office in a nation that was not her own.[12]

It's not surprising that these tensions later propelled Rudolphina towards the political philosophy of one of Theodor Herzl's main disciples, Martin Buber—a Bohemian, Central European, Jewish intellectual who emphasized the importance of personal relationships and whose notion of Jewishness was simple but profound: "To be a Jew means to be human and to be it in the Jewish form!"[13] Buber developed a more mystical notion of Zionism that was simultaneously both socialist and universalist, which appealed deeply to Rudolphina. Buber's utopian Zionism imagined a multicultural Jewish homeland in which Jews lived in harmony with Muslims and Christians in Palestine. Coexistence based on a celebration of difference, rather than a denial and erasure of it, would form the foundation of Buber's new Zion, a vision that made profound sense to Rudolphina. Buber taught Rudolphina on a theoretical level what she learned from the Glockels on a personal level: how to temper her Zionism with a deeper devotion to universalism. She came to reject the chauvinistic assertion that Jewish particularity meant Jewish superiority—a notion that ran through some other strains of Zionism. The Glockels' generosity, magnanimity, and personal example made that kind of Zionism not only provincial, but anathema to Rudolphina. Ultimately, it was Buber's brand of Zionism and the Glockels' vision of social democracy that converged to lay the foundation for Rudolphina's worldview.

Rudolphina began her studies in the notoriously rigorous faculty of Natural Sciences at the University of Vienna in 1909 and as a university student found meaningful outlets for her Zionist fervor. It was during these years at the university that she helped found the Viennese branch of the Jewish scouting movement *Blau-Weiss*, a purposefully Zionist organization that demanded its members speak Hebrew, commit to an ethical way of life, and connect to nature by taking strenuous hikes through the Austrian countryside.

Rudolphina vividly recounts these energetic excursions in which she and her Zionist comrades hiked through friendly Alpine villages, waved their blue and white flags, sang Hebrew songs, and danced around campfires. She described these adventures as magical and formative ones during which her ideals and experiences were in blissful harmony. Some of her dear friends from *Blau-Weiss* would soon die fighting for Austria in World War I. Others would die fighting for Palestine years later. Many of those who survived became lifelong friends.

While at university, Rudolphina's passion for Zionism also led her to seek membership in the serious and exclusive Theodor Herzl Zionist College Students Club. Traditionally a male-only organization, Rudolphina insisted on joining despite loud objections, particularly from a young medical student from Teplice (near Prague) named Rudolph Menzel. Subsequent debates between them about the future of the Jewish people and long hikes in the hills on the outskirts of Vienna must have served as an intoxicating aphrodisiac; the two were soon engaged.[14] Rudolphina went on to complete her PhD dissertation on the sugar compounds in urine in 1914 and published two scientific papers with her advisor Ernst Zerner on their discovery of pentose compounds in the urine of patients suffering from a rare metabolic disease.[15] Not long after World War I erupted, she and Rudolph were married.[16]

1915–1938: Becoming the Menzels

Rudolphina omitted any detailed descriptions of World War I from her memoir but noted that as soon as the war began, Rudolph enlisted in the Austro-Hungarian army as a medic. Rudolphina went to work as an assistant professor at the Viennese Institute for Cancer Research, and as the war wore on, she was appointed manager of the chemical laboratory in a large factory outside Vienna. In 1917, Rudolph was stationed at a garrison hospital in the small pastoral town of Linz, 175 miles west of Vienna. Not long afterwards, Rudolphina decided to quit her job and move to Linz so she and Rudolph could finally live together after two years of wartime separation.[17]

The couple found a room in a small Linzer inn where they were looked after by a kindly innkeeper who cooked their meals and sat with them while they ate. They were amused to discover they were the first Jews she had ever met, but they shouldn't have been. Linz, Hitler's childhood home, was a provincial Austrian city in 1917 and home to only around nine hundred Jews out of a population of ninety thousand. The Jewish community there was close-knit, comprised primarily of small business entrepreneurs and shopkeepers.[18]

In those final months of the war, when she and Rudolph were in their mid-twenties but still essentially newlyweds, Rudolphina did something very uncharacteristic: she relaxed. She read poetry, took short hikes around Linz, and enjoyed evening strolls with Rudolph along the banks of the Danube. If the Menzels had ever planned to have children, this would have been a good time to do so; however, it is unclear whether their childlessness was by choice or by fate.

When the war came to a tumultuous end, fear of looters and lawlessness briefly overcame Linz as soldiers straggled back from the front lines. One of the units of Rudolph's hospital had been serving as a kind of de facto prison for wounded soldiers, some of whom were deserters, foreign nationals, or criminals. Despite intense pressure to keep these patients locked up, Rudolph played a decisive role in the humanitarian decision to discharge them.[19] Once order was restored in Linz and the hospital was secured and reopened, most doctors returned to the homes they had left before the war. Rudolph, by contrast, was unwilling to abandon his patients.

Even though they had planned to return to Vienna after the war, the Menzels made the unusual decision to settle in Linz permanently. Leaving behind the lively cultural and intellectual world of Vienna was a monumental move, particularly for Rudolphina. It meant abandoning her professional career as a chemist, saying goodbye to her dear friends in *Blau-Weiss*, and no longer living near the Glockels. But it also gave the young couple a fresh start, which for Rudolphina meant establishing an adult life with comfortable distance between her and her despised stepmother.

Postwar conditions in provincial Linz and its environs were bleak; being on the losing side of the war meant widespread austerity and hardship. But Rudolph's salary as a doctor was dependable, and his kindness and attentiveness helped him quickly become a well-known and well-liked doctor in town. By 1919, Rudolphina had established herself as something of a local personality as well; she was remembered as a "funny, lively little character with a gravelly voice" who often rode around town on a bike.[20]

The young couple reached out to the Linz branch of the Austrian Social Democratic Party, then known as the German Social Democratic Party of Austria, and quickly became infamous for their refusal to join the party as "Germans." They insisted on joining as members of the

The Extraordinary Life of Rudolphina Menzel

Jewish nation: the same steadfast position that Rudolphina had been cultivating since her years arguing with the Glockels before the war in Vienna. Rudolphina became active in the party's campaigns to establish workers' libraries and old age homes and to improve healthcare in Linz.[21] It didn't take long before she was appointed chair of her district's Freie Schule Association, following in Frau Glockel's footsteps.

The Menzels also cultivated connections with local Jewish groups, few of which were sufficiently Zionist in their opinions. To remedy this, Rudolphina decided to regularly deliver public lectures on Zionism, and Rudolph ran for the board of the Linz Jewish community on a Zionist platform, a position he won and held for many years.[22]

It was around this time that Rudolphina adopted her first dog. In her memoir, she underplayed this life-changing event, noting dryly that "after the war, I was given the opportunity to fulfill a childhood dream and keep dogs."[23] The only account of this transformative encounter was provided by the Austrian veterinarian Joseph Bodingbauer, a renowned breeder of Doberman Pinschers and a lifelong friend of the Menzels who later became an expert in small animal dentistry.[24]

Bodingbauer reported that shortly after the war ended, he sought treatment for his lung disease at a small hospital in Linz. For reasons he didn't explain, he apparently brought his prized Doberman Pinscher Pazzos-Naxos with him.[25] A young doctor named Rudolph Menzel happened to be on duty that day, and after examining Bodingbauer and asking questions about his condition, Rudolph befriended Pazzos-Naxos as well. Noticing Rudolph's keen interest in the dog, Bodingbauer offered to arrange a "private demonstration" for the doctor and his wife to watch Pazzos-Naxos perform his latest tricks. Bodingbauer provided no details about the actual meeting. He only reported that something about this private demonstration "prompted the Menzel couple to devote themselves to cynology."[26] Perhaps Rudolphina's scientific imagination was ignited by watching the intense, communicative relationship between Bodingbauer and Pazzos-Naxos. Perhaps the experience prompted her to wonder how this well-trained dog thought, learned, and remembered. Or perhaps she simply longed for a dependent creature to take care of because she and Rudolph were childless. Bodingbauer simply noted that shortly afterwards, the Menzels asked him to help them get their own Doberman Pinscher.

The little Doberman puppy Bodingbauer first gave the Menzels died of distemper after only a few days, but the replacement Boxer puppy he found for them at the von Donnhauf Kennel in Linz fared much better. He was a robust, brindle-colored male with a sturdy physique and the gangly, adorable silliness common to Boxer puppies. Rudolphina was immediately besotted with her new young charge and named him Mowgli, presumably after the main character in *The Jungle Book* by Rudyard Kipling.

For someone who knew nothing about dogs, learning how to care for and train one was not simple—particularly since Rudolphina was barely five feet tall, and her Mowgli quickly grew into a very muscular, rambunctious dog. To learn the latest techniques, Rudolphina read the classic German-language dog-training manual written by Colonel Konrad Most. Published in 1911, Most's manual was recognized not only as the bible of dog training at the time but for many decades afterwards.[27] Most advocated for the use of "forced inducements," including choke collars, spiked collars, and whipping, to produce the desired behaviors in dogs. His methods relied on the theory of "operant conditioning": that by teaching a dog to associate failure to complete a task with punishment and success with reward, a dog's behavior can be effectively shaped and controlled (B. F. Skinner conducted experiments on humans relying on the same theory over thirty years later). Rudolphina mastered Most's theories of canine motivation and, being naturally authoritative, seemed to have a knack for applying his training techniques as well; Mowgli quickly became a very obedient dog.

Training Mowgli to sit, stay, and come and watching him learn to obey her commands sparked Rudolphina's competitiveness. She soon decided she wasn't content to just stroll around Linz holding her well-trained dog on a leash like other provincial pet owners; she wanted to enter Mowgli in dog shows. In his first outing at the Munich Dog Show on September 18, 1921, Mowgli came in sixth.[28] The judge commended Rudolphina on her excellent handling and noted that Mowgli obeyed her instructions with lightning speed; he also remarked on the strange language she used with her dog, for Rudolphina trained Mowgli only to respond to commands in Hebrew. While it must have been difficult for Rudolphina, who was accustomed to being first in her class, to come in sixth with Mowgli, she was undaunted by her mediocre debut. She took the disappointment as a challenge, and it was not long before Mowgli went on to become an outstanding show dog.

Rudolphina could have enjoyed her new hobby the way other bourgeois Austrian housewives did, as a fun opportunity to socialize with likeminded people while traveling around Central Europe to participate in leisurely canine competitions. She could have gone a step further by joining her regional breed club and helping to organize dogs shows, popular pursuits for more energetic dog show enthusiasts. Or at a respectable extreme, she could have started to breed Mowgli at home by selecting a suitable bitch for him to mate with and enjoying the challenge of raising a litter of puppies.

Alternatively, as an intellectual, she could have done what the great German novelist Thomas Mann did when he got a dog: write a philosophical book about her relationship with Mowgli the way Mann wrote a memoir about his German short-haired pointer Bashan in 1919.

> I am happy to interrupt my literary occupation in order to speak and play with Bashan. I repeat—to speak with him. And what do I find to say? Well, the conversation is usually limited to repeating his name to him—his name—those two syllables which concern him more than all others, since they designate nothing but himself, and thus have an electrifying effect on his entire being. I thus stir and fire his consciousness of his ego by abjuring him in different tones and in different degrees of emphasis to consider the fact that he is called Bashan and that he is Bashan.[29]

While perhaps it wouldn't have been as lyrical as the Nobel Prize-winning Mann's, Rudolphina's book could have provided a nice, modest outlet for her musings on the inner life of her dog and the nature of canine subjectivity.

Instead, Rudolphina decided she would not be content to simply own, train, show, or ruminate about the inner lives of dogs; she wanted to become a cynologist and systematically breed and study them. For her, keeping dogs was not going to be a bourgeois hobby but a serious professional undertaking that enabled her to investigate scientific questions and solve societal problems. Mowgli had inadvertently provided a gateway to an entirely new career.[30]

If Rudolphina wanted to become a cynologist, Rudolph was going to build a kennel that enabled her to do so. He started looking for a house with a large yard and with Bodingbauer's help soon found a two-story villa on a nice plot of land not far from where Bodingbauer himself lived in the small, industrial village of Kleinmunchen just outside of Linz. Surrounded by workers' cottages, the setting was semirural and picturesque, and with their socialist sympathies, it suited the Menzels perfectly; they liked the idea of living amongst the proletariat. The Menzels contracted some manual laborers from the neighborhood to construct a set of dog houses in the yard, and they moved in on May 16, 1922. Rudolphina was thirty-one years old, Rudolph was thirty-three, and Kleinmunchen would be their home for the next sixteen years.

The Menzels were unlikely denizens of the little, rural Austrian town, not just as Jews but as bourgeois, cosmopolitan Jews. Rudolphina knew their new neighbors perceived them as "quirky, rich, childless people" who didn't know what to do with their time or money.[31] But Rudolph routinely treated their neighbors' minor medical problems, and Rudolphina provided jobs for many of the local youth as kennel assistants and household helpers, which gradually ingratiated them into the neighborhood.

Once settled in Kleinmunchen, Rudolph made the short commute into Linz every day to his medical office. At home, Rudolphina began to learn all the practical skills necessary to feed, house, mate, whelp, nurture, exercise, vaccinate, and otherwise care for dogs and puppies. Through connections she had developed while exhibiting Mowgli at regional dog shows, she began purchasing particularly outstanding Boxer bitches to mate with him, and it wasn't long before the little huts in her Kleinmunchen yard were filled with dozens of noisy Boxers. Fortunately, there were only two houses nearby, the nearest occupied by an old lady who was deaf. This may have helped quell any complaints about the cacophony of barking dogs that emanated constantly from the kennel, which eventually housed up to sixty dogs at a time.[32]

In addition to mastering the practical aspects of kennel-keeping, Rudolphina began to consider all the exciting questions that were then percolating in the scientific world of cynology. Many of these concerned the ancient origins of dogs, the processes of domestication, and the perceptual abilities of the dog, but others mirrored questions that were being considered in human psychology more broadly, such as whether behavior was rooted in biological instincts and determined by genetic inheritance or shaped by social environment and individual experience. Like her fellow Austrian Jew Sigmund Freud, who was enthralled with exploring questions about the origins and mutability of behavior amongst humans, Rudolphina became fascinated with investigating similar questions amongst dogs.

Most of the scientific research that had been conducted on dogs up to that point took place in experimental laboratories where dogs were used as disposable research subjects, such as the laboratory Russian physiologist Ivan Pavlov built to study the roots of canine behavior and classical conditioning. Rudolphina had a different plan. She wanted to observe domestic dogs in their "natural" environment, a household-like habitat in which they formed bonds with people. Rather than being experimented on and disposed of once her observations were complete, her research subjects would be socialized, trained for specific tasks, and sold to reliable owners. Her canine research program, therefore, would serve as both a scientific laboratory and a functioning kennel.

With her PhD in chemistry and as former director of a state chemistry lab, Rudolphina knew how to construct experiments, test hypotheses, manage staff, and run a laboratory, but launching this new endeavor demanded that she transpose these skills onto a completely new field of scientific inquiry. This required an intelligence flexible enough to master an unfamiliar body of theoretical literature and an imagination prodigious enough to design original experiments using novel methods, not to mention incredible and unwavering perseverance—particularly when you decide, as Rudolphina did, to study primary development, one of the most challenging areas in animal behavior.

To begin, Rudolphina designed an extraordinarily ambitious longitudinal research study in order to compare and evaluate the early development of hundreds of Boxer puppies bred over successive generations. She created a comprehensive system to weigh, measure, and observe each puppy every day from the moment of birth in order to record their growth and keep track of developmental variations in their behavior as they matured. By taking notes on daily changes in their mobility, reflexes, sensory development, feeding, excretion, and sociability, the study's design enabled her to focus intensely on the period from birth to maturity, which is a span of about ten months for a dog.

She documented when each newborn puppy started suckling and with what degree of energy and enthusiasm. She kept track of when each puppy raised its head and lifted its tail, both vertically, horizontally, and in a wagging motion. She recorded the ages at which puppies locomoted with their front legs and with their back legs and observed that while some newborns run earlier and some later, by the third week of life, all run normally. She logged when each puppy started to defecate in the typical crouch position (usually around fourteen days). She also made meticulous observations about when later behaviors emerged, such as male leg-lifting to urinate (as early as sixty-seven days) and mounting behavior (around eight weeks). She went on to identify behaviors that were subject to environmental conditions. For example, she postulated that temperature fluctuations, inadequate access to nutrition, the onset of illness, or being removed from the litter could delay the emergence of a puppy's playing behavior (normally observable by the fifteenth or sixteenth day of life).[33]

In short, Rudolphina collected an extraordinarily large amount of data that enabled her to make a series of bold conjectures about the differences between inherited and acquired individual traits. While basic instincts like smelling, suckling, and defecating were present at birth, she observed how other traits like friendliness or aggressiveness emerged differentially and how they were shaped by each puppy's environment and experience. She concluded that canine behavior was determined by a complex interplay between the puppy's inherited mental traits, their physical condition, and the variety and intensity of environmental stimuli the puppy received at particular moments in its early life. By showing that there was considerable malleability in canine temperaments, she demonstrated how both intended and unintended intrusions in a puppy's environment at critical developmental stages affect a puppy's subsequent behavior as an adult dog, regardless of their genetic inheritance.

She began to classify her Boxer research subjects by temperament type, using the same terminology Ivan Pavlov had borrowed from Hippocrates to describe human personalities; she described most as "choleric"(-excitable), some as "sanguine" (lively), few as "phlegmatic" (calm), and only a handful as "melancholic" (inhibited).[34] She also borrowed from Freud's theory that human childhood could be divided in to five distinct stages—oral, anal, phallic, latent, and genital— by advancing a similar theory that dog development could also be divided into five stages: the "vegetative period" (birth to two weeks), the "awakening period" (three to six weeks), the "entry-into-the-world" period (seven to sixteen weeks), "prepuberty" (sixteen to twenty-four weeks), and "puberty" (twenty-four to forty weeks). By advancing this hypothesis, she became the first to posit a comparative ontogeny between children and puppies, thereby providing a paradigm of canine development that has shaped the field ever since.[35] Moreover, her prescient interest in the relationships between early trauma, memory, and later behavior addressed a set of theoretical questions that have continued to occupy researchers in ethology, psychology, and zoology.

Despite the fact that Rudolphina's pioneering research was recognized as groundbreaking at the time, it has since been largely forgotten or ignored, particularly in the subsequent literature in English. Part of the reason for this may be due to the fact that it was never translated from the original German.[36] But American researchers clearly knew of her work because it is often mentioned or alluded to in scholarly publications in the field. For example, in 1950 the American geneticist and comparative psychologist John Paul Scott (writing with social psychologist Mary Vesta Marston) directly acknowledged her research: "The best recent studies of the development of behavior in puppies have been done by Baege and the two Menzels. The former studied a litter of five puppies... Menzel's [sic] studies on the other hand were done on a large population of purebred Boxers."[37] Rather than explicitly credit Rudolphina for developing the research design and providing the theoretical foundation for their subsequent work, however, Scott and Marston write: "It will be seen that these periods in general coincide with those of which we have used."[38] "Coincide" is too weak a word to describe the overlap. Scott and Marston simply appropriated the categories Rudolphina had created and published many years earlier.

Scott's failure to adequately acknowledge the influence of Rudolphina's work persisted at least until 1965, when he (writing with biologist John L. Fuller) published the classic *Genetics and the Social Behavior of the Dog*, which quickly became known as the definitive study on genetic and environmental influences on the behavior of dogs and the most important scientific treatise on the subject in the English language.[39] In fact, some of Scott and Fuller's hugely influential findings were based on methods and conclusions borrowed directly from Rudolphina's pioneering research. While Scott and Fuller included a reference to "the Menzels" in their bibliography, they did not cite the work directly nor did they credit it explicitly in any way.[40] Their failure to adequately acknowledge Rudolphina's work is made more curious by the fact that they sustained a lively correspondence with her throughout the 1950s. Perhaps these omissions were more careless than deliberate, but Scott and Fuller made no effort to correct them as they became known as the pioneering researchers in the field.[41] Adding to this erasure, in other scholarly references to Rudolphina's research, Rudolph is often the only one credited.[42]

The data from Rudolphina's longitudinal research on puppy development would be collected cumulatively during the 1920s and '30s, but she simultaneously developed a variety of short-term studies and practical initiatives. Rudolphina didn't specify the exact date that Bodingbauer introduced her to Emil Hauck, a prominent Viennese veterinarian who was the breed warden of the Austro-Hungarian Police Dog Association, an expert on the ancient origins of dogs, and a renowned breeder of Austrian Pinschers and English Bull Terriers.[43] It is clear, however, that by the time she started her kennel operations, Hauck had become not only a good friend but a mentor as well. Likely due to a combination of Hauck's influence and her socialist convictions, Rudolphina became fascinated with the idea of breeding and training dogs that actively contributed to modern society.

This was a different kind of service than that which dogs had long provided to humans in primarily agrarian settings, where for thousands of years they served as shepherds, sled dogs, guard dogs, and hunting companions. Instead, modern service dogs were being raised and trained to perform specific tasks alongside men and machines in urban, industrial societies. These modern notions about the dog as a kind of canine citizen-soldier only accelerated after World War I, during which British, American, and German armies used hundreds of trained dogs to carry messages, sniff out enemies, and attack on command.[44] This was a time when service dogs were also being trained to assist in nation-building and colonial regime control more broadly.[45]

Although Rudolphina had been introduced to Boxers entirely by chance, it turned out the breed was well-suited to police work, one of the most vibrant areas of modern service-dog training.[46] Tasks at which Boxers excelled included following scent tracks to locate missing people, identifying crime scene evidence, guarding prisoners, and attacking criminals fleeing from or attacking police officers. The whole concept of using a dog to track and catch criminals had, in fact, originated not far from Linz in 1896 when the Austro-Hungarian criminologist Hans Gustav Adolph Gross suggested using dogs as police "assistants."[47]

In order to breed Boxers to be effective police dogs, Rudolphina had to apply her understandings of Mendelian genetics and the principles of heredity to the concrete task of selectively mating dogs. This required her—like all dog pedigree breeders—to effectively practice a kind of lay eugenics since the foundational pillar of dog breeding rests on the eugenic notion that reproduction should be engineered to produce particular outcomes.

It is important to note that Rudolphina was not unaware of the troubling similarities between the eugenic beliefs that formed the basis of the pragmatic world of dog breeding and the ways they later became transmogrified into the guiding political ideology of the Nazis (an overlap made more pronounced by the fact that the word for breed and race are the same in German: *rasse*).⁴⁸ She made no excuses for this similarity—Rudolphina openly credited the Nazi scientist Friedrich Alverdes with inventing the mathematical formula that best expressed her "scientific world view." His formula stated that behavior is the function of two factors; one constant (inherited genetic material) and one variable (environmental influence).⁴⁹ She recognized that this same equation provided the foundation for the Nazis' "blood and soil" theory but noted that "when you're a scientist, you can't choose your beliefs."⁵⁰ She did, however, make it clear that there must always be a "wall" between science and politics.⁵¹

Becoming a lay eugenicist in order to engineer auspicious pairings between her Boxers was not difficult for Rudolphina; it simply meant mastering complicated, though fairly straightforward, information that she either already knew, could learn independently, or could acquire from her friends Bodingbauer and Hauck or from other established breeders. Rudolphina's efforts began to exceed and diverge from established practices when she decided to make her breeding program more efficient by developing a system that could reliably identify, predict, and breed for specific canine character traits. This was a considerably more ambitious undertaking which required her to master the voluminous scientific literature on the variability of temperaments and the processes of heritability, burgeoning areas of scientific inquiry that were of enormous interest not just to cynologists but to psychologists, biologists, and eugenicists as well.

On a theoretical level, such a system would allow her to collect valuable data on the patterns of inheritance of particular character traits amongst a controlled population, enabling her to contribute directly to ongoing scientific debates about the genetic origins of behavior by using dogs as research subjects.⁵² On a practical level, learning how to breed for particular traits like pugnacity, protectiveness, and loyalty would make her kennel more efficient at producing Boxers with carefully selected dispositions, ensuring that time and energy were not wasted raising police dogs with a limited innate ability to perform the tasks that were expected of them.

Since Rudolphina's breeding program was folded into her longitudinal research study, she was in an ideal position to notice the enormous variability amongst puppies born from the same parents and weaned under identical circumstances. She observed that puppies in the same litter not only exhibited obvious morphological differences, they also possessed unique personalities. If she could develop a tool to identify these individual character traits, she would be able to better predict the aptitudes they would possess as adult dogs. Such a tool would then provide both a means and a validation for breeding dogs with specific temperaments, thereby revolutionizing the ways dogs were bred, raised, and trained.

Perhaps inspired by the personality tests developed by psychologists during World War I to identify soldiers prone to nervous breakdowns (and later adapted for use in modern industry to create more efficient workplaces by screening out potentially problematic employees), Rudolphina set out to create a kind of psychological assessment for dogs.[53] She developed a battery of "tests" which she used to test dogs' responses to a variety of stimuli in order to rate them on nine "fundamental attributes" that combined to create a dog's particular temperament.[54]

Information gleaned from these tests could also be applied to later breeding as well. By calibrating each dog's "score" with an understanding of which traits were dominant, recessive, or heterozygous, she could then coordinate reproductive pairings that would reinforce desired psychological attributes and produce puppies more likely to possess particular aptitudes. In order to further refine her efforts, Rudolphina tested dogs between the ages of six and nine months of age, since younger dogs by definition have less experience with the world and are therefore more likely to reveal traits that were inherited rather than acquired. Of course, she had to temper her findings with the knowledge that dogs—and especially young dogs—can display different traits at different times depending on whether they are sick, tired, hungry, or overly distracted.

Interestingly, Rudolphina did not design an IQ test to rate the overall intelligence of a dog. Canine intelligence, and the capacity of a dog to learn from experience, were not at that time understood to be a fundamental attribute one could test for, nor was it thought to be particularly indicative of trainability. Similarly, she did not develop a test to determine the willingness of a dog to exhibit their identified aptitudes. Quantifying a dog's intelligence and motivation engaged a separate set of psychological questions about an individual dog's attitude towards learning. These qualities could best be observed and addressed in the framework of the dog's training program and often only in the context of the individual dog's relationship with their trainer, where the desire to please by following commands was most evident. By combining observations of a dog's eagerness to learn with the results of the temperament test, Rudolphina

The Extraordinary Life of Rudolphina Menzel

could further refine the process of matching a dog to the appropriate form of service, be it guarding, tracking, attacking, or some combination of all three. Rudolphina repeatedly emphasized that in order to ensure good results, the right dog had to be assigned to the right job.[55]

It is worth reproducing the contents of two tests Rudolphina published in her first scientific paper on canine temperament testing in 1932.[56] Here, she illustrates the specific situational tests she developed as well as the particular observations she made for two test subjects who displayed markedly different temperaments: a seven-month-old male named Thanatos and a nine-month-old female named Suata.

THANATOS, SEVEN-MONTH-OLD MALE

Tracing Disposition

Does the dog trace with his nose or with his eyes? *Especially eager and yet intensive tracing without any visual correction.*

Random traces or directed traces? *Random traces. (Three corners).*

On or off the leash? *On the leash.*

How does the dog behave when an object with an unfamiliar scent is placed close to an object belonging to his owner? *The first time he took his master's object without paying any attention to the other object; when repeated, he sniffed both objects and took the one belonging to his master.*

Temperament, Tractability, Fetching Disposition, and Playfulness

Does the dog trust his master? (Does he bare his teeth, make gentle, playful swats with his paws, etc.?) *Absolute trust.*

Does the dog like to play? *Very much.*

Are his playful nips forceful? *Very forceful.*

Does he lock his teeth when biting? *Yes.*

How does the dog relate to fetching? (Lack of interest, sniffing the object, taking hold of the object?) *Very enthusiastically, he happily picks the object up and plays with it.*

Is the dog attentive to his master and does he follow him willingly? *Very attentive, follows happily.*

Is it easy, difficult, or impossible to divert him from his master? *He is not at all diverted from his master.*

How does the dog overcome obstacles that come between himself and his master? *Quickly and skillfully.*

Acuity, Attentiveness, Reliability of Behavior

Behavior toward harmless strangers? *Indifferent.*
Behavior toward friendly strangers? *Trusting, but somewhat reserved.*
Behavior toward threatening strangers (also when threatened with a stick)? *Very energetic reaction, becomes immediately aggressive.*
Behavior in light of unusual phenomena

Acoustical

Commotion behind curtains or covers? *Very attentive, barking.*
Commotion aimed against the dog? *He immediately runs away.*
Gunshots? *Attentive, slightly aggressive.*

Optical

Opening an umbrella? *Immediately aggressive.*
Billowing sheets or cloths? *Immediately aggressive.*

Other conspicuous or random observations?

Alertness: see above.

Defensive Instincts, Combativeness, Tenacity

How does the dog behave in light of an attack?
Against his master? *Very energetically, he immediately engages and holds fast.*
Against himself? *Very energetically, he immediately engages and holds fast.*
Does he defend immediately, after a while, or not at all? *Immediately.*
If not at all, does he consider the situation a game or is he afraid? _____
Does he run away or does he remain at a respectful distance? _____
Does he bark or not? _____
In a fight:
How does the dog react to a stick? *Very resolutely.*
Does he endure blows? *He bears strong blows without intimidation.*
How does he react to screams? *Very well, he endures them.*
In a particularly aggressive fight:
"Henze test" with a stick: *Very resolutely, does not flinch or retreat.*
Note: *Also passed the "Henze test" with gunshots resolutely.*
Observations concerning the dog's physical development: *Especially strong development.*
Evaluation of the dog's fundamental attributes:

1. Courage +++
2. Defensive Instincts +++
3. Pugnacity +++
4. Acuity +++
5. Temperament +++
6. Endurance +++
7. Tractability +++
8. Fetching Disposition +++
9. Tracing Enthusiasm +++

Overall evaluation: *Excellent disposition.*

SUATA, NINE-MONTH-OLD FEMALE
Tracing Disposition

Does the dog trace with her nose or with her eyes? *Rather indifferent tracing, very distracted, is much more interested in the dogs playing in her vicinity.*

Random traces or directed traces: *Little interest in random traces. A bit better with directed traces, but transitory and with nose held high as a result of too strong an interest in that which is to be found.*

On or off the leash? *On the leash.*

How does the dog behave when an object with an unfamiliar scent is placed close to an object belonging to her owner? *This test wasn't performed because of a lack of interest on the part of the dog.*

Temperament, Tractability, Fetching Disposition, and Playfulness

Does the dog trust her master? *Yes.*

Does the dog like to play? *Seems to be distracted, loses interest in the other dogs.*

Are her playful nips forceful? *Not very, she's too distracted.*

Does she lock her teeth when biting? *No.*

How does the dog relate to fetching? *Interested, but strongly distracted by the other dogs.*

Is the dog attentive to her master and does she follow him willingly? *Not very attentive and only moderately obedient.*

Is it easy, difficult, or impossible to divert her from her master? *She is not at all diverted from her master.*

How does the dog overcome obstacles that come between herself and her master? *Very skillfully, very engaged.*

Acuity, Attentiveness, Reliability of Behavior

Behavior toward harmless strangers? *Friendly and trusting.*
Behavior toward friendly strangers? *Friendly and trusting.*
Behavior toward threatening strangers (also when threatened with a stick)? *Relatively easily intimidated, anxious.*
Behavior in light of unusual phenomena

Acoustical

Commotion behind curtains or covers? *Attentive.*
Commotion aimed against the dog? *Anxious, is easily driven away.*
Gunshots? *Attentive.*

Optical

Opening an umbrella? *Somewhat anxious.*
Billowing sheets or cloths? *Somewhat reserved.*

Other conspicuous or random observations?

Alertness: _____

Defensive Instincts, Combativeness, Tenacity

How does the dog behave in light of an attack?
Against her master? *She wants to protect him, but hesitates to jump into the fray.*
Against herself? *Anxious.*
Does she defend immediately, after a while, or not at all? *Not at all.*
If not at all, does she consider the situation a game or is she afraid? *She is afraid.*
Does she run away or does she remain at a respectful distance? *Remains at a distance, but is ever ready to flee.*
Does she bark or not? *She does not bark.*
In a fight:
How does the dog react to a stick? *Did not perform the test.*
Does she endure blows? *Did not perform the test.*
How does she react to screams? *Did not perform the test.*
In a particularly aggressive fight:

"Henze test" with a stick? *Did not perform the test.*
Notes: *Isolated and pampered animal, essentially still immature, not nervous, but languid.*
Observations concerning the dog's physical development: *Strong development.*
Evaluation of the dog's fundamental attributes:

1. Courage –
2. Defensive Instincts ++
3. Pugnacity 0
4. Acuity ?
5. Temperament +
6. Endurance *very weak*
7. Tractability + –
8. Fetching Disposition +
9. Tracing enthusiasm + –

Overall evaluation: *Failed the test.*

With her publication of these techniques and outcomes, Rudolphina was among the first to effectively transpose the field of personality testing from human psychology to dog breeding and training. Canine temperament testing has since become standard practice around the world, not just for evaluating the dispositions of pedigree puppies and service dogs but for evaluating the temperaments of stray and rescued dogs as well.[57] While Rudolphina has received some acknowledgment for her foundational research in these areas in later German-language publications, her contributions to the invention of canine temperament testing have been almost entirely absent from English-language literature.[58]

Her application of these techniques to her breeding program in Kleinmunchen drastically improved her outcomes, and her "Linzer Boxers" soon became known as an outstanding line of obedient, strong, and courageous dogs with excellent aptitudes for tracking work, making them much sought after by Austrian, Swiss, and German police and military units. Bodingbauer soon began applying her techniques to improve outcomes in his Doberman Pinscher breeding program and no doubt others followed suit.[59]

As it turned out, Central European police dog enthusiasts weren't just interested in Rudolphina's methods and her dogs; they quickly became interested in Rudolphina herself. By the late 1920s, she was traveling all over Central Europe, giving lectures on canine cognition and demonstrating the utility of her pioneering temperament testing methods at municipal police departments, military bases, and cynological associations. She also began making regular trips to Berlin as a canine consultant for both the criminal investigation department of the Berlin police and the German Reichswehr's central military dog-training facility.[60] In fact, she was such a frequent visitor there that a special residential pavilion was constructed solely for her use to ensure she was as comfortable as possible while in residence.[61]

When the petite, and by then plump, Rudolphina arrived in Berlin, she was wined and dined and treated like royalty by her Aryan hosts. While women were generally prohibited from riding in official military vehicles (a rule no doubt intended to keep officers from joyriding with their wives), Rudolphina proudly noted that she was driven to and from the Canine Research Department of the Army High Command in a sleek, black Mercedes-Benz. She was nonplussed by the special treatment and reasoned that "if Marie Curie had visited Germany, she too would have been driven around in a company car!"[62] By then she had developed a close friendship with the Department's director, the revered Colonel Konrad Most, the author of the renowned German dog-training bible, which no doubt contributed to the cult-like status she acquired in the German-speaking world of police-dog training.

Rudolphina's pioneering research was not limited to canine psychology or development. She also became deeply engaged in all manner of problems concerning animal behavior and physiology and was soon publishing a whole range of scientific studies. Her research on canine olfaction resulted in a landmark 1930 publication on the dog's sensory physiology and mental ability to process information.[63] She proved that the canine sense of smell is so acute and discriminating that, with the right training, a dog can learn to recognize a person's individual scent— even if the person had only touched an object for a short time and their scent had been overlaid with other odors. Rudolphina's background in chemistry was foundational to this work. She demonstrated that the most significant source of an individual's scent came from skin secretions and described the three biological processes that contribute to the development of human scent: perspiration, sweat secretion, and secretions from the subcutaneous sebaceous glands. Then, using her acute understanding of the ways different odors break down over time, she invented a complex method for training dogs to recognize a person's scent at different time intervals and under varying conditions. Rudolphina's research effectively revolutionized the way dogs were used for tracking criminals and changed the way the forensic utility of police dogs was understood. Her research was particularly important for proving the reliability of the evidence dogs provide by using their sense of smell. Rudolphina was soon known not only as an excellent police-dog breeder but also as one of Europe's leading cynological researchers on animal behavior.

In 1930, Rudolphina organized the first Convention of the Dog-Breeding Authority of the Austrian Association of Cynologists, which came to be known as a "landmark in Germany cynology."[64] The most eminent cynologists and military dog trainers in Central Europe converged on Linz for the event. Following the conference, Rudolphina was awarded the "golden medal of honor" by the Austrian Boxer Club in recognition of her extraordinary contributions to the field, and the "Linz Resolutions" were issued, which outlined the best practices for training tracking dogs.[65]

By this point, the Menzels' small villa in Kleinmunchen had become something of an international meeting place for a diverse assortment of overwhelmingly gentile dog enthusiasts. For some, visiting the Menzels was an adventurous excursion to the countryside in order to commune with intellectually engaging eccentrics.[66] For others, it was a pilgrimage to discuss the latest ethological research with some of Europe's most famous cynologists. Great lovers of good conversation, the Menzels were winning hosts, and discussions about the ancient origins of dogs, the sociology of wolf packs, the heritability of character traits, and the evolutionary processes of canine domestication lasted long into the night.

Approaching the Menzels' house was an unforgettable experience for many of their visitors. The thunderous barking of dozens of Boxers reverberated throughout the neighborhood when anyone walked down their street. A sign on the Menzels' gate warned visitors to beware of attack dogs, but those bold and curious enough to venture inside received a warm and lively Menzel welcome. Their salon was filled with books and dog show trophies, and guests gathered around coffee and cake served by one of their devoted maids. One guest, a journalist named Friedrich Feuchtinger, provided a particularly vivid memory of his visit:

I ring the doorbell of a friendly villa near the perimeter of the extensive meadow and farmland, and soon thereafter, I find myself in a snug reception room, where a display cabinet filled with badges, sculptures, medals, and laurel wreaths indicates to me that I am in the right place. While I reverently admire all the prize trophies from the various dog shows all over the world, the door opens and in front of me stands Dr. Rudolphine Menzel in her winning simplicity, and soon a conversation is flowing. Convincing words, aglow with deepest love for our faithful housemate, the dog, warmly reach my ears and open up to me a new, hitherto quite foreign field. How irresistibly one feels the truth of the assertion of psychologists who say that people who demonstrate love of animals also completely fill their place in society, for the Menzels more than others always have a judicious understanding of the hardship and misery of the present time and innumerable people can testify to their generosity and their readiness to help.[67]

Feuchtinger reported with astonishment how Rudolphina's Boxers could follow new and old scent tracks even after rainfall or when covered by snow. He couldn't believe how her dogs could reliably identify clothes that had been stored in glass containers and kept for many weeks. He concluded that "the Boxer especially is predestined to be man's most reliable helper and that he stands in the first line of combat dogs as a valuable breed. The many beautiful prizes that have been awarded to this kennel prove that the Boxer compound in Kleinmunchen plays a leading role."[68]

Rudolphina recounted a lively variety of visitors to Kleinmunchen, including a high-ranking Prussian officer, the director of the Austrian Gendarmes, a university professor from Honolulu, a museum director from Switzerland, and other well-known painters, writers, and scholars. These dog fanciers came to the Menzels' villa by car, motorcycle, or streetcar; one day, a group of Jewish workers from Palestine even arrived on bikes on their way from Trieste.[69] Some guests were elegantly dressed, while others were not. Some were aristocrats, others vagabonds. Socialist Rudolphina proudly reported that "when they were with us, all differences disappeared."[70]

Joseph Bodingbauer, Konrad Most, and Emil Hauck were regular visitors, as were the Swiss scientist Eugen Seiferle and the world-famous Austrian ethologist Konrad Lorenz (who later became a Nazi and won the Nobel Prize). Though the Menzels exchanged letters with Ivan Pavlov, the great Soviet expert on canine physiology, by the early 1930s he was too elderly to visit them in Kleinmunchen.[71]

Even the Chief Rabbi of Vienna paid a visit to the Menzels in Kleinmunchen—an occasion that was later vividly recalled by Rudolph:

Many years ago Dr. Zevi Hrisch Chayes, the well-known scholar and Chief Rabbi of Vienna, visited the small Jewish community in Linz on the Danube. The small Jewish community, an insignificant minority in the flourishing city of the Austrian Alpine provinces, had not much to offer by way of objects of interest. To make up for this deficiency, at the afternoon reception given by one of the notables, two "freaks" were introduced to him in the persons of my wife and myself. For so we were regarded by our fellow Jews, our way of life and work being exceedingly strange, not to say exotic, in their eyes.

To keep a pack of sixty hounds, to devote the whole of your leisure, to say nothing of a considerable amount of money, to train these dogs in a lot of weird tricks, such as to follow policemen, gendarmes, and hunters on Sundays and holidays from the crack of dawn to late at night over fields and moors, in the broiling sun or winter cold—these goings on could have been understood if there had been any proof in them. But to waste money on such a hobby, moreover to insist on it as work of Zionist importance, was too much for our fellow Jews. The kindest among them considered it a more or less harmless pastime and smiled good-naturedly at any mention of the subject. But Chayes understood without much explanation. He realized that this type of work was at least as valuable as activities to impress outsiders or to make a name in the scientific world. He did not laugh when we were introduced to him, or utter the customary formal phrases. On the contrary, when he learned of the nature of our work his eyes lit up and he exclaimed: "Our Jewish renaissance expresses itself in marvelous ways!"[72]

It perhaps goes without saying that both Menzels were quite eccentric figures in the German-speaking dog world of the 1920s and '30s: a bastion of aging, former-imperial aristocrats and gentile petit-bourgeoisie (though many of them shared the youthful experience of having participated in nationalistic hiking groups—albeit with very different ideological underpinnings).[73] These cynophiles came together in mutual enthusiasm for the then still relatively new craze of modern dog breeding and training; many of them eagerly became Nazis after Hitler rose to power.[74] Rudolphina in particular stood out in these crowds as a short, lively Jewish woman from the intellectual elite of Vienna; she was not only a serious scientist but an unrepentant Zionist to boot.

The Menzels were not the only Jews in interwar, Central European dog sports, but they were among the very few. Max Hilzheimer (1877–1946), a German-Jewish zoologist and expert on domestic animals in antiquity, was a prominent cynologist at the time. Joseph Schwabacher (1880–1953) started breeding dogs around 1915 at his influential Secretainerie Kennel and continued to produce outstanding German

Shepherds and Schnauzers until he was forced to flee to England in 1939; he also translated Max von Stephanitz's classic 1923 treatise on the German Shepherd Dog into English. Ernest Loeb (1909–2004), another German Jew, was a respected breeder of German Shepherds and was instrumental in establishing the breed in the United States after he fled Nazi Germany in the mid-1930s. In the US, he was known as "Mr. German Shepherd."

Certainly, the Menzels' Jewishness had never been a secret amongst their dog breeding friends. Rudolphina trained all her dogs only to respond to commands in Hebrew, which meant that the gentile police officers working with them had to learn Hebrew commands. While it was common practice to train police dogs in a foreign language so criminals could not communicate with dogs that were pursuing or attacking them, the spectacle of German military and police-dog trainers barking instructions at their dogs in rudimentary Hebrew must have provided Rudolphina with endless amusement and great satisfaction. She no doubt derived pleasure from the threatening, demonic name she gave to her kennel as well: she chose to call it *B'nei Satan*, Hebrew for "Children of Satan."

The Menzels often went out of their way to make sure their guests, hosts, publishers, and colleagues knew that they were Jewish in order to ensure that "no embarrassing situations" arose should their identity be unexpectedly exposed in a way that would pose a problem for them.[75] Yet, it seemed that desire for the Menzels' company and for Rudolphina's canine expertise worked to override any latent or overt antisemitism in her circles.

Indeed, the Menzels' Jewishness did not prevent the Austrian Cynology Association from insisting that they represent Austria at both the third International Cynological Congress held in Nazi Germany in 1935 and the fourth International Cynological Congress held in Paris in 1937, nor did it preclude them from holding high office in any of the major canine organizations in Austria, Germany, or throughout Europe. The Menzels were members of both the Austrian and German Canine Research Boards, honorary expert advisers to the Austrian Rural Police and other police forces, and official judges of form and service appointed by the *Federation Cynologique Internationale*, the venerated European federation of national kennel clubs founded in 1911. Further, being Jewish did not prevent Rudolphina from serving as the Chief Breed Warden and Chief Performance Manager of the Austrian Boxer Club or from serving later as the Executive Director of its Upper Austria branch. In fact, she was repeatedly offered the role of chairman of these organizations, but she refused due to the same conviction that precluded her from assuming a leadership role in the Austrian Social Democratic Party—she felt that someone from the Jewish nation should not hold a prominent and representative position in a national organization that was not their own.[76]

* * * * *

If a Jew has a dog, either the dog is no dog or the Jew is no Jew.
YIDDISH PROVERB

Here in the *Eretz*, our people found their way back to the earth and it is time we found our way back to the dog. Working with dogs is pioneer work, and this is why we approach you, the upcoming generation. Help us reclaim the dog for our people. He who has not overcome the fear of the dog from the ghetto is not a renewed Jew...we call on the renewed Jew. We call you. Make room for a new pioneering path, to reclaim the dog for the building of our country.[77]
RUDOLPHINA MENZEL, 1943

As Rudolphina's canine activities expanded throughout the 1920s, the Jewish organizations of Linz, former beneficiaries of Rudolphina's considerable energy and organizational leadership, must have wondered what had happened to her.[78] Why was she wasting her time breeding Boxers in Kleinmunchen, hosting cynological conferences, and lecturing to police departments in Berlin? To them, it seemed her interests and commitments had shifted dramatically and with them, her social circles. But Rudolphina had not abandoned the Jewish community or her deep-seated Jewish commitments. On the contrary, from her earliest days as a breeder of Boxers, she was strategizing about how best to combine her passion for dogs with her Zionist convictions. In fact, when she wasn't conducting trailblazing canine research, innovating training methods, breeding outstanding Boxers, or traveling all over Central Europe as a cynological consultant, Rudolphina spent the 1920s continually writing letters to leaders of the Jewish settlement in Palestine (the Yishuv).[79] Her goal was to persuade Zionist leaders to invest in training and using dogs as a vital part of their state-building project.

Rudolphina was not the first to recognize that dogs could be of service to the Zionist cause. In 1913, the Zionist leader Arthur Ruppin brought two Doberman Pinschers to Palestine to serve with *Hashomer*, the Jewish defense organization that had been founded in 1909. The outbreak of World War I stymied Ruppin's plans to import more guard dogs from Europe, but when Rudolphina began her campaign to establish service-dog training courses in Palestine, he became one of her strongest advocates.[80]

Like Ruppin, Rudolphina knew Palestine was a perilous place for Jewish settlers, and she believed that dogs could contribute as valuable "workers" by serving Zionist purposes in myriad ways that were going completely unrealized. For the kind of agrarian economy Zionists were building, herding dogs could be of obvious help, but modern service dogs could serve the interests of the Yishuv in more critical ways as well. Guard and patrol dogs could protect the perimeters of remote Jewish settlements, military camps, and ammunition stores by using their sharp senses of smell and hearing to report the approach of intruders.[81] Patrol dogs could also provide effective modes of security for newly planted fields, orchards, and forests by protecting them from arson and sabotage, frequent methods of Arab resistance to Zionist settlement. These kinds of service dogs were not only more effective and reliable sentinels than humans, but using them in these ways also freed up limited manpower to perform other tasks.[82]

In addition, messenger dogs could maintain vital links between otherwise isolated outposts by carrying messages through hostile areas that might endanger humans—communication that could provide a real strategic advantage when roads were blocked or tensions flared. Transport dogs could carry emergency medical supplies or ammunition to isolated combatants during conflicts, particularly over rugged terrain. Mine-detecting dogs could be trained to recognize the smell of subtle disturbances in the earth used to cover land mines; this was important information, as using homemade mines to destroy Jewish-owned vehicles became a more prevalent anti-Zionist tactic in the 1930s.[83]

Interestingly, Rudolphina did not initially describe using dogs as offensive weapons against the Arabs. Rather, she described ways they could be deployed in defensive roles, mainly to protect Jewish land and save Jewish lives. This is not to suggest that she harbored any particularly sympathetic attitudes towards Palestine's Arabs or that she was morally opposed to training dogs to harm them—like most Zionists of her day, she did not give much thought to the consequences of Zionist settlement for the native Arab populations. Moreover, as an expert breeder of police dogs, she certainly knew how to train dogs to perform lethal attacks and disable designated enemies. But Rudolphina knew that as colonial subjects of the British Mandate, Jews in Palestine were prohibited from owning weapons—and trained attack dogs would certainly qualify as such.[84]

Persuading Zionists to recognize the value of guard and patrol dogs was not difficult, but convincing them to actually invest in training and caring for dogs presented a unique challenge. Jewish ambivalence towards dogs was deeply rooted.[85] Not only did Jews traditionally avoid dogs due to Biblical and Talmudic dictates about their uncleanliness but Diaspora Jews also had a long history of fearing dogs given that gentiles, particularly in Europe, often used dogs to chase, attack, and terrorize them. Moreover, most of the leaders of the Yishuv came from the Jewish ghettoes of Eastern Europe, where dogs were generally shunned, avoided, or both. In addition, because many Zionist pioneers fashioned themselves as socialist revolutionaries, they regarded pet-keeping as a bourgeois affectation entirely out of keeping with the pioneering ethos. To them, caring for dogs was not only a habit of non-Jews but a habit of non-Zionists as well, practiced by the kind of urban, affluent Jews from Central and Western Europe who were their ideological opposites. Rudolphina therefore had to be careful to differentiate between companion dogs, which could be perceived as meaningless pets, and trained service dogs, which were vital, utilitarian workers. While the former were superfluous to the Zionist project, the latter were imperative to it.

Resolving the multilayered Jewish cultural resistance to dogs posed one set of challenges for Rudolphina, while the logistical requirements for establishing a completely new modern canine infrastructure in Palestine posed another. To properly care for, breed, and train service dogs, dog handlers would need to be recruited and taught the fundamentals of veterinary health, the principles of genetics, and the basics of animal behavior and psychology. They would require instruction in how to appropriately stimulate and socialize puppies and how to train dogs to dependably perform specific tasks. This meant that Hebrew-language dog-training curricula had to be developed, dog-training courses had to be organized, and a whole variety of practical arrangements had to be made. Leashes, collars, harnesses, muzzles, kennels, and dog food would need to be imported or manufactured. Reliable veterinary care would also need to be available, which was particularly imperative in Palestine's hot, desert climate where canine diseases were rampant. Most importantly, a dependable population of trainable dogs would have to be imported, bred, or both. All of this required money, time, and canine expertise—resources that were scarce in British Mandatory Palestine in the 1920s, where Jewish knowledge about and affection for dogs was in extremely short supply and where resources to properly care for and train them was almost nonexistent.

Rudolphina recognized these overwhelming obstacles but remained resolute. By 1927, she managed to convince the Jewish Agency in Palestine to send two young kibbutzniks to Kleinmunchen to learn how to train guard dogs. After a few weeks of intensive tutorials, they returned to Palestine with a male/female pair of Rudolphina's Boxers and detailed instructions about how to breed and train more.[86] Rudolphina also began to coordinate with Zionist youth groups in Central Europe so that whenever they sent groups of young people to Palestine, she made sure they brought trained Boxer dogs with them.[87]

Finally in 1932, after over a decade of waging her letter-writing campaign, Rudolphina received a personal invitation from Yitzhak Ben-Zvi, then chairman of the national executive committee of the Yishuv and later the third president of Israel, and Yaakov Pat, commander of the Yishuv's militia, to come to Palestine herself in order to teach members of the Haganah how to train and use dogs.[88] The invitation was hand-delivered by a female emissary sent specially from Palestine to meet Rudolphina at her home in Austria—an indication, perhaps, of the Jewish Agency's growing understanding that Rudolphina's plan for establishing a canine corps of trained service dogs could make a critical contribution to Zionist settlement, making her important enough to merit an official invitation.

The Jewish Agency coordinated the logistics and made all the arrangements for Rudolphina's trip: they reserved venues for training courses, recruited prospective dog handlers, arranged meetings with various officials, and organized all transportation and lodging. By early 1934, Rudolphina's complicated itinerary had been finalized, and she was ready and excited to depart.[89] At the last minute, however, her trip was postponed. Political tensions in Austria spiked suddenly in January 1934 after the arrest and execution of an Austrian worker named Peter Strauss by the fascist Dollfuss government. This event triggered a series of protests that culminated in a brief but dramatic civil war in February 1934, when Austrian Social Democrats rose up to challenge Dollfuss' authoritarianism and were immediately crushed by government forces. Over 1,500 people were killed in riots against the police and army, and many associated with social-democratic institutions were arrested including Rudolph, who spent the night in a local jail.[90] One of the policemen who came to arrest Rudolph and search the Menzels' home for opposition party material was an acquaintance from the kennel club who had often been their guest. As a result, the polite and perfunctory search ended with a friendly chat about new dog-training methods.

Rudolph was placed in one of the finest cells and received excellent treatment, which Rudolphina attributed to the fact that she was so admired and respected by the local police for her contributions to police-dog training. The Menzels' dear friends the Glockels were not so lucky; as leaders of the Austrian Social Democratic Party in Vienna, Leopoldine was imprisoned for six weeks in miserable conditions, and Otto was sent to one of the first concentration camps. Rudolph's involvement with the Social Democratic Party did not go unpunished after the uprising, however. He lost his job as a referee at the Arbitration Court for the Disabled and his position as the head of the Child and Marriage Guidance Agency in Linz, which resulted in a substantial loss of income for the Menzels.[91]

These events shook Rudolphina deeply.[92] When Hitler became chancellor of Germany a year earlier, her response had been phlegmatic. She wrote that nothing really changed for her and Rudolph and that their gentile friends had reassured them "again and again" that what was happening in Germany would be impossible in Austria and that everything they heard about the treatment of Jews there was grossly exaggerated.[93] But after the events of February 1934, she saw how Nazi ideas were gaining traction amongst the dispossessed and marginalized in Austria and knew the widespread unemployment and broader economic stagnation would only fuel the rise of authoritarianism there.

In response to the events, and no doubt in reaction to the Austrofascists' use of military dogs trained according to her methods, Rudolphina ended her professional relationships with the Austrian and German police departments, stopped giving lectures on dog behavior and training, and refused to write articles for the German-language press in her field of expertise.[94] She had long been aware that dogs could be misused for the wrong purposes and had always been against the gratuitous use of police dogs to inflict harm. In fact, Rudolphina harbored a deep ambivalence about the societal role of the police in general. In her memoir, she explicitly articulated this dilemma from a socialist perspective:

> One's heart bursts when one thinks that the police dog almost without exception comes into play against the weak in the state, against the little evildoer, never against the big and never against the official villains. I express it bluntly that in many cases my heart is absolutely on the side of the evildoer and not on the side of the accuser. After all, in the end every evil that the police dog is supposed to help uncover is mostly a consequence of the prevailing social conditions. Whoever sits in a villa doesn't need to break into it.[95]

In a strained rationalization offered to reconcile these contradictions, she went on to explain that "if the dog were a means of punishment, according to our *weltanschauung*, we would have to reject promoting its use... but it is a means of research, an instrument with which one can help someone who is not protected as well as unmask someone who is protected... Everything in the world can be misused. With dynamite you can build roads and tunnels or you can use it to commit human murder on a large scale... Dynamite cannot be held responsible for such abuse, and neither can the people who work with it."[96]

Once the Austro-fascists had taken over in 1934, however, she could no longer tolerate or reconcile these tensions and found it impossible to countenance further involvement with the new state's police force. It is interesting to note that Rudolphina cast her decision to end her professional relationships with these various law enforcement agencies as one she made on her own. She made no mention of the subsequent Nuremberg Laws that forbade Jews from having dogs in Germany or her fear of similar laws being imposed in Austria.

In the months after the brief Austrian civil war, Rudolphina's eagerness to undertake her long-awaited and long-planned trip to Palestine became only more acute. She wrote:

> In the fall of 1934 I went to Palestine for the first time. This trip had been planned for many years and was finally scheduled for March 1934. But of course, due to the February events, I did not travel. For many years already, I had made efforts from Europe to organize the service dog system, which was infinitely important for the safety of the young colonial country. I had trained people as service dog handlers at my place and was in constant contact with the Land [Palestine]. But finally, I had to get things going personally.[97]

The Extraordinary Life of Rudolphina Menzel

She arrived in the Haifa port on October 1, 1934 to find a boisterous, diverse, and rapidly changing country. Still overwhelmingly populated by Arab Muslims and Christians (almost 900,000 according to the British Mandate census), there were almost 300,000 Jews in Palestine at the time, and thousands were arriving each month from an increasingly antisemitic Europe. She was met at the port by Rose Vitales, a fundraiser for Jewish causes in Palestine, and Maurice Beck Hexter, the Director of the Palestine Emergency Fund and member of the Jewish Agency Executive Committee; they had arranged a very busy month for her, packed with meetings, lectures, and dog-training courses.[98]

Rudolphina's first stop was Mt. Scopus in Jerusalem, where a day-long meeting had been arranged with the Hebrew philologist Naphtali Herz Tur-Sinai, professor of Hebrew language at Hebrew University and first president of the Academy of the Hebrew Language. Rudolphina sought out Tur-Sinai's expert advice and assistance in order to write the first dog-training manual in Modern Hebrew, a complicated philological task since Modern Hebrew had not yet invented a vocabulary for canine training and care. Individual words for particular forms of equipment, such as "muzzle," "harness," "leash," "protective suit," and "kennel" needed to be crafted out of ancient Hebrew, as did terminology for dog training and dog behavior, such as "visual sign" and "stimulus linkage."[99] She and Tur-Sinai spent the whole day together, working through dog-training terms and concepts line by line to make sure her manual was as accurate and accessible to readers as possible. Konrad Most's famous German-language dog-training manual clearly served as a starting point, but the subsequent Hebrew-language manual illustrated how Rudolphina had since come to reject Most's punitive operant conditioning methods and to view inflicting harsh punishment on dogs when they failed to perform tasks on command as counterproductive.

Though the recently founded university sat on a hot, rocky hilltop and was composed of only a few dusty, concrete buildings, Professor Tur-Sinai routinely wore a European-tailored, double-breasted suit to work. Rudolphina, just off the boat from Austria, was no doubt similarly attired in a sensible wool blazer and skirt. The two German-speaking intellectuals must have looked as if they were attending a poetry reading in Vienna rather than sitting in a hot, concrete room on the edge of the Judean desert translating a German dog-training manual into Hebrew: one of many surreal tableaus that exemplified Rudolphina's entire trip.

From Mt. Scopus, she embarked on a whirlwind series of meetings with a veritable who's who of prominent Yishuv leaders, including Yitzhak Ben-Zvi, Yaakov Pat, and Moshe Sharett, the new director of the Jewish Agency's political department (and later the second prime minister of Israel). An experienced and convincing speaker, albeit in a heavily German-accented Hebrew, Rudolphina detailed the myriad benefits that trained service dogs could provide for guarding Jewish settlements and protecting Jewish lives. Winning over these senior officials was not easy; there were many competing demands on the Jewish Agency's time and attention, resources were extremely limited, and her cause was still seen by many as peripheral, impractical, or both.

The main goal of her trip was to lead three multiday dog-training courses for members of the Haganah at Tel Mond, Kibbutz Yagur, and Mikve Israel in order to build up a cadre of certified and confident dog handlers. She taught her students how to train guard dogs to identify, frighten, and pursue intruders, burglars, and thieves. She also gave evening lectures to general audiences about the value of trained service dogs, performed dog tricks for children, and met with the wives of British Mandate officers to talk about Boxers.[100]

One of her other goals for the trip was to assess the public health challenges facing her canine agenda. While the British Mandate authorities had constructed a public healthcare apparatus as part of their colonial regime, there were still repeated outbreaks of rabies and persistent problems with parasites, tick-borne illnesses, and other canine diseases.[101] In her meeting with the veterinarian Dr. K. S. Krikorian, an official in the Mandate's Department of Health and an expert on the incidence of rabies, Rudolphina inquired about the frequency of vaccination campaigns, the efficacy of quarantines, and the uses of strychnine poisoning to control outbreaks; she was also curious about the practicality of periodic programs to cull the local jackal population, which were often vectors for canine disease.

An equally pressing objective was to evaluate the quality of the extant dog population in Palestine in order to identify suitable dogs for training. Importing pedigree dogs from Europe had proven to be impractical. The Boxers she had sent previously had either failed to function effectively in the intense Middle Eastern heat or had quickly succumbed to the numerous unchecked canine diseases in the country. Rudolphina knew she had to find a way to establish and maintain a reliable population of trainable dogs. She later made it clear how dismayed she was by the local canine conditions she found on her trip:

The Extraordinary Life of Rudolphina Menzel

On the paths and roads... neglected dogs swarmed and ran about, mostly semi-wild dogs, just like in the villages of the Arab fellah. The local people had no knowledge of how to keep, take care, and use dogs. The dogs were so full of ticks it is almost impossible to describe—from the dogs' ears hung not just clusters but entire vineyards of ticks—according to the ironic words of one of the farmers. A large portion of the animals had skin diseases, and when humans approached them some of them turned their backs, while others simply laid down apathetically.[102]

The dog population of Palestine was indeed an eclectic and often sickly hodgepodge at the time.[103] There were a few pet dogs: terriers, spaniels, mongrels, and others, usually cast-offs or descendants of purebred dogs owned by the wives of British Mandate officers or recent immigrants from Western Europe. There were scattered groups of small hounds that had been imported by British officers for recreational hunting, a diversion that ebbed and flowed during the Mandate period depending on the amount of interest shown in the sport by the individual officers stationed there and the level of tension in the country (by the time Captain Geoffrey Warden imported a group of fox hounds in 1932, biweekly hunts near Ramle were routine).[104] There were elusive populations of salukis, traditional Arab hunting hounds who were objects of careful Bedouin breeding for clandestine sport and coursing (salukis have long been celebrated as hunting companions in Arabic poetry, their lineages memorized by their devoted Bedouin custodians and their courage and beauty revered). But the overwhelming majority of dogs in Palestine were the indigenous, semiwild dogs that Rudolphina observed scavenging in garbage dumps and roaming freely outside towns and villages. Sometimes these dogs were captured by Bedouin and chained outside their tents to serve as canine alarm systems, but otherwise they were despised, feared, ignored, or avoided.

Despite their miserable conditions, these free-ranging Palestinian dogs were of great interest to Rudolphina for two reasons; first, once properly bred, selected, and cared for, she reasoned they could function as a limitless reservoir of local, climate-adapted canine candidates for her training programs. Second, Rudolphina viewed them as ideal research subjects. She recognized them as a local variety of "pariah dogs"—a term traditionally used by cynologists to describe ownerless, semidomesticated or feral dogs, which are found all over the developing world.[105] These dogs offered an ideal prospect, she wrote to her cynological mentor Emil Hauck, to study the missing link between wolves and dogs and to observe canine domestication in action. "Do it before it's too late" he advised her, before increasing colonization brought more European breeds to the area who would interbreed with and dilute these "living fossils."[106]

Rudolphina's ethological research program would have to wait, however; her immediate concern was whether these dogs could be trained and used to breed other trainable dogs. Due to her extensive research on the heritability of character traits and her experience conducting canine personality tests, she knew that any kind of dog, regardless of provenance, could possess traits that would make them good candidates for training. She therefore set out to conduct temperament tests on as many of these free-ranging dogs as she could, hoping to identify at least a few who were cooperative, sociable, and responsive. Rounding up and evaluating feral dogs was not simple, however, and there is no record of how many dogs she managed to conduct temperament tests on during this trip.

Before it was time to return to Austria, Rudolphina managed to squeeze in a performance of Leopold Lindtberg's Hebrew translation of Friedrich Wolf's 1933 play "Professor Mannheim" at the Habimah Theater in Tel Aviv, visit the Kinneret, and stop by the potash works at the Dead Sea. While she may have been discouraged by the canine state of affairs in Palestine, she was delighted by the progress of the Zionist movement and later recorded her rosy first impressions of the country she had fantasized about since her youth:

> I found a free people, free despite limits on immigration and the like, found myself really rooted in the land, as I had dreamed. It was true after all, my people did not consist of religious Jews and coffeehouse-sitters. It looked like I had seen it in my dreams. Intellectuals turned farmers, proud workers, a beautiful, healthy youth. I saw with my own eyes the beginnings of new forms of human cohabitation, I saw modest beginnings of a new humanity. Full of hope and recuperated, I returned to Europe after having attended to my work. I held my head up twice as high, I had regained my faith in humanity.[107]

As she boarded the boat in Haifa, she was pleased by what she had accomplished even if she had been unable to secure any firm or formal commitments for her dog-training plans. She left behind a small cadre of dog trainers to continue her work, expanded her network of supporters amongst the upper echelons of the Jewish Agency, and made a few more individual converts to her cause. She had also been able to personally assess the canine conditions in the country, which meant she could strategize more concretely about the overwhelming amount of work that lay before her.

A few months after her return home to Kleinmunchen, Rudolphina and Rudolph boarded a train to Nazi Germany as part of the Austrian Delegation at the third International Cynological Congress. Held in April 1935 in Frankfurt, the congress included delegates from thirty-two countries and was followed by a dog show that welcomed over sixty thousand people and featured 3,300 dogs.[108] The theme of the conference was "the inheritance of the physical and mental characteristics of the dog"; the relationship between heredity and social behavior was of course of great interest to Nazis at the time.

The opening dinner was held at Frankfurt City Hall, a majestic, medieval building with a neogothic façade that had once been the site of German imperial coronations. As Rudolphina anticipated, the Nazis understood the event, like the Olympics held in Berlin the following year, as an opportunity to showcase German "greatness" on an international stage. She and Rudolph sat dutifully through the dinner surrounded by swastikas while their jovial German colleagues, with whom they had been breeding and training dogs for years, privately admitted to them how appalled they were by the new government. Later, when Hans von Tschammer-Osten, the newly appointed Director of Sports in the Third Reich, tried to grab her arm, Rudolphina deftly pushed another guest between them and disappeared into the crowd; she had no interest in exchanging pleasantries with the Nazi official.[109] She was careful to observe the reactions of other national representatives and noted their symbolic connotations in light of later political developments. For example, the leader of the French delegation "an old, internationally known cynologist and proud aristocrat...kept a cold distance from everything that was German and treated every attempt at approach from the other side with cool, relaxed, and impenetrable courtesy."[110] But when von Tshammer-Osten sidled up to him with what Rudolphina described as a "lovingly interested expression," the unapproachable Frenchman quickly softened and the "ice was broken...from then on there were good relations between France and Germany...von Tschammer-Osten had effectively won them over."[111] The English delegation, by contrast, was apparently immune to von Tschammer-Osten's charms—they chose to leave the hall altogether rather than shake his hand. Rudolphina wryly concluded that the "English resistance proved to be more durable than the French."[112]

The Menzels found the entire display of "Byzantine patriotism" highly distasteful and chose to manifest their outrage over the whole affair by using their keynote to directly contradict Nazi dogma about the "immutability of germinal substance" and the intractability of genetic differences between races:[113]

> In terms of modern genetics, we understand race to be the sum of all typical traits, physical and emotional, or more specifically, their genetic predisposition, which are passed on according to a certain norm of reaction. Race is what is passed on. Accordingly, strictly speaking, every human being, every animal has their own race. This new, simple concept of race differs essentially from earlier, often hard to understand concepts of race, which should be used with caution.[114]

In other words, the Menzels argued that race was effectively synonymous with individual heredity and did not provide a basis for dividing and classifying human beings by groups. As opposed to the multiple races the Nazis were so invested in classifying and ordering based on imagined and fixed genetic differences, the Menzels openly declared that there was only one human race, made up of individuals each with a unique mosaic of genetic traits that were particular to them, not to their *rasse*. Any attempt to suggest otherwise was entirely misguided from a scientific point of view.

No doubt due to their heretical and brazen rejection of the Nazi's belief in biologically distinct races, the editors of *Der Hund*, the official publication for German dog sports, refused to publish the Menzels' scientific article in the commemorative edition following the congress, despite having been a regular outlet for reports on their research. A picture of the Menzels captioned "the German researcher couple from Austria" was nevertheless printed in the publication right next to a picture of Hitler and Goring.[115] It is unclear whether the Menzels stayed for the dog show following the conference, at which Hitler's pet German Shepherd unsurprisingly won first place.[116]

The Menzels returned to Kleinmunchen after the spectacle in Frankfurt and got straight back to work. Rudolphina reported that she and Rudolph still didn't feel any particular urgency to leave Europe. Though they had seen the dark clouds of fascism gathering since the Austrian civil war, they were used to unsettled and agitated times, they were deeply absorbed in their various projects, and they decided to keep their distance from politics.[117] Of this time, she noted that "when I was in my little house, far from the city, immersed in my work, I felt as though I were in a 'glass house,' far away from all the ugliness of daily life."[118]

The Extraordinary Life of Rudolphina Menzel 43

By 1936, however, the Menzels had become embroiled in a neighborhood feud that gradually escalated in to a legal battle.[119] The political atmosphere in Austria and the attendant rise in antisemitism had finally come to a head in a very sharp and personal way. It began with a letter from a new neighbor complaining about the howling and barking coming from the Menzels' kennel. Rudolph's initial reply to the neighbor was polite: he suggested the gentleman must have known he was buying a house next to a kennel when he purchased the property, and "it was impossible to demand total quiet in an open development, since a great variety of noises were inevitable. Naturally, no promise could be made that the dogs would not bark ever again." But soon it became apparent that barking and howling—and other problems that had long been tolerated in the neighborhood, like dogs periodically escaping into neighboring yards, the smell of rotting dog food, and the constant swarms of flies generated by the kennel—were being catalogued and logged with the local authorities. As the conflict escalated, Rudolph's tone changed markedly as he parried with the neighbor. In a letter he submitted to the municipal court in 1936, Rudolph wrote persuasive, though somewhat patronizing, rebuttals to each charge. First, he explained that the kennel served lofty scientific goals:

> Maintaining our dog kennel is not a superfluous hobby but a serious labor of many years, which serves very far-reaching scientific research purposes. Last year at the International Cynological Congress, we were able to publish comprehensive statistics on the heredity of mental skills. This is accompanied by research into the performance capacity of dogs in the service of humanity. This work is a matter of public interest. Especially before the forum of a police authority, the public interest of such endeavors does not need to be specially emphasized...We are already forced to notice a large number of impairments to our scientific work. Nobody can demand of us to destroy the work of a lifetime.

Second, he explained that the noise complaints were overblown since the Boxer is not a noisy breed:

> In principle, the Boxer is a quiet dog which does not yap or bark without a reason. At night, the largest part of our dogs are [sic] looked after and locked up in the house or in the kennel, and only a small number of watch dogs (2–3) are free in the run.

Third, regarding the escaped dogs, Rudolph made it clear that they had replaced their wire fence with a stronger, wooden one. And finally, in an elitist turn, he proffered his psychoanalytic opinion that the complainant was ill-advisedly displacing anger about his own life. He then rattled off the names of local police authorities who were the Menzels' personal friends and visitors, all of whom they of course knew from their work training police dogs.

The matter was ultimately referred to the provincial court, a hearing was held, and a series of directives were issued to the Menzels. They were ordered to secure proper permits for the kennels (which had been built fourteen years earlier without proper permitting). They were required to securely seal all dog food to prevent rat infestations, and they became subject to inspections to ensure that they regularly cleaned the entire property to reduce the pervasive smells that evidently emanated from it. As a result, the Menzels had to devote considerable time to applications for kennel construction permits and attending to fixes for all the other "violations."

This unfortunate and perhaps inevitable neighborhood drama unfolded at the same time that Rudolphina was feverishly working to conclude her monumental longitudinal research study on canine development and busily refining her plans for the creation of an entirely new canine infrastructure in Palestine. By this point, the Jewish Agency was also clamoring for Rudolphina to return as soon as possible and wanted her to remain in the country for at least a year.[120] Since the Arab Revolt of 1936, dogs had proven their value for guarding and patrolling, and an increasing number of Jewish volunteers were eager to learn how to become dog handlers. But apart from a short trip to Switzerland, where they spent a week as guests of the Swiss Cynological Society, the Menzels spent the mid-1930s close to home, immersed in their work and fending off complaints from their neighbors in Kleinmunchen.[121]

It wasn't until spring 1937 that Rudolphina finally returned to Palestine, this time with Rudolph, for an intensive two-month trip. They were still not prepared to emigrate for good; Rudolphina's longitudinal research study was not yet completed, Rudolph's medical practice was busy, and their kennel—though under attack from their neighbors—was still full of Boxers. On their way to Palestine, they stopped in Vienna to visit Leopoldine Glockel, who was by then elderly, frail, and sick; it was the last time Rudolphina saw her beloved friend.

The Menzels disembarked at Haifa Port on April 26, 1937 and were driven directly to Jerusalem for a working meeting to finalize their itinerary. The exhaustive schedule of activities the Jewish Agency arranged for

The Extraordinary Life of Rudolphina Menzel 45

Rudolphina indicated the seriousness with which they had come to view her dog-training enterprise. The Jewish Agency had recruited cadres of male and female students to attend three intensive, multi-week courses: two for guard-dog handlers and one for tracking-dog handlers.[122] The courses were rigorous. Rudolphina demanded that her students not only master a complicated set of practical training skills but a raft of scholarly material as well; she required them to understand how dogs think, the historical origins of dogs, the evolutionary processes of domestication, the inheritance of mental traits, and the different ways dogs have been used throughout human history. Final examinations tested the students' practical skills as well as their theoretical understanding, and only those who mastered both were recognized as licensed dog handlers. Rudolphina selected particularly outstanding graduates to participate in her follow-up teacher-training course, which was aimed at producing a corps of certified dog-training instructors.[123]

In addition to teaching the demanding dog-training courses, Rudolphina's schedule was packed with public lectures for general audiences all over the country. After these talks, she was invariably bombarded with questions about dog care and training, reflecting an increasing appreciation for the usefulness of dogs and a growing desire to provide adequate and appropriate care for them. She also delivered academic presentations on animal psychology for more scholarly audiences at the Hebrew University in Jerusalem, met with officers from the Jewish Agency, and convinced members of the underground Irgun Militia to buy puppies and start socializing and training them as patrol and guard dogs.

Rudolphina also held a variety of meetings with representatives from various British government agencies. She met with the Mandate government's chief veterinary officers to discuss the status of anti-rabies campaigns. She met with the Palestine police force at their kennels on Mt. Scopus to give advice on the latest training methods for tracking suspects and pursuing criminals.[124] She also found time to establish the Palestine Kennel Club in Jerusalem in order to provide oversight for the breeding and training of dogs.[125]

As she travelled around the country, Rudolphina made frequent stops to conduct canine temperament tests on potential candidates for dog training. In a rough diary she kept of her trip, she noted that she conducted temperament tests on almost two hundred dogs.[126] Any free time was spent working feverishly on her Hebrew-language dog-training manual so it could be sent to the printer before she returned to Austria.

Save for a brief visit to show Rudolph the Dead Sea, there had been no time for sightseeing.

As a Zionist, Rudolphina was delighted to see how much the Jewish cities, towns, and villages had grown since her first visit three years earlier, but as a cynologist, she was again dismayed by the canine conditions she encountered. Despite the combined efforts of university researchers and Jewish and British veterinarians, there were still frequent outbreaks of rabies, constant problems with tick and flea infestations, and persistent troubles with other canine diseases. Moreover, there was seldom enough money to properly feed, shelter, and care for dogs, so even the best dogs were often short-lived.

While some small kennels had been established and were producing occasional litters of purebred dogs for sale (usually Terriers, Pointers, Spaniels, Boxers, Alsatians, and Dachshunds), many of these purebreds were allowed to roam freely and mate indiscriminately, often with the local street dogs, producing puppies who all-too-often inherited the fearful, nervous traits of their semiwild parent and were unsuitable for training.[127] The people running these kennels were often inexperienced novices who lacked proper knowledge about canine reproduction. Bitches were frequently overbred, puppies were improperly weaned, and breeders didn't know how to effectively stimulate and nurture puppies in the first days and weeks of life, socialization that was essential to their positive cognitive development. The irregular collection of dogs that was being used for guarding and patrolling had been randomly assigned to these tasks without being tested to determine if they had the temperaments necessary to succeed at them, making for an assortment of often unreliable and unpredictable guard and patrol dogs. Rudolphina knew that importing purebred dogs from Europe was still not a viable solution. She witnessed a disaster unfold at Kibbutz Sha'ar Hanegev during her trip when twenty-one purebred German Shepherds, which had been imported at great expense for a dog-training program, got infected with a local strain of flea cholera as soon as they arrived in the country. Not only did all the dogs die from the disease, their human caretaker did as well.[128]

Despite the entreaties of Jewish Agency officials who implored them to stay longer, the Menzels set sail from Haifa on June 30, 1937, exhausted and uncertain about the challenges that lay ahead. Rudolphina's assessment of the trip was both terse and grandiose: she noted only that she

The Extraordinary Life of Rudolphina Menzel

successfully managed to lay "the organizational groundwork for the entire Jewish service dog system in the Land."[129]

Once back in Kleinmunchen, the Menzels had barely enough time to unpack before they boarded a train for Paris, where Rudolph had been invited to deliver the keynote address on behalf of the Austrian delegation at the fourth International Cynological Congress, which was being held as part of the 1937 Paris World's Fair. This sprawling showcase for the world's scientific and technological achievements took place during a time of escalating political crises all over Europe; the theme for the canine exhibition was "the psychology of the domestic dog."[130]

The Menzels had never been to Paris but considered themselves spiritual descendants of the enlightenment ideas that were born there. As they strolled along the banks of the Seine arm in arm, the city of light worked its magic on the middle-aged couple, bestowing a rare respite from their increasingly untenable life in Kleinmunchen and the mountain of practical problems they left behind in Palestine. While preparing for the events of the congress, the Menzels relaxed and appreciated what were to be their last peaceful moments in Europe.[131]

In a show of Zionist pride, Rudolph made a point of beginning his keynote lecture by noting that he and his wife had just returned from a wonderful trip to Palestine. He then went on to describe the findings that represented the culmination of Rudolphina's pioneering sixteen-year longitudinal research study on canine behavioral development, based on her detailed, daily inspections of over one thousand individual Boxer puppies bred over sixteen years and eight generations. At the conclusion of the congress and in recognition of the achievement, the Menzels were awarded a medal by the city of Paris.[132] The research was published later that year as *Welpe und Umwelt* (Puppy and Environment). It was the first study to comprehensively evaluate changes in development amongst a large population of blood-related dogs over an extended period of time while paying particular attention to how the character of the mother dog affected canine development. It also examined how the intensity and duration of external environmental influences affect puppies during the first three months of life.[133]

The Menzels returned from the cynological congress in Paris to a very anxious Austria. Rudolph continued to see patients and skirmish with the neighbors over the legal permits for their kennel. Rudolphina kept busy with her numerous projects and resumed hosting distinguished visitors including an Egyptian cynological consultant from the Cairo police.[134] She was still on very good terms with the local Boxer club, which in January 1938 presented her with a polished crystal trophy bearing an inscription in her honor written by one of Linz's most well-known Nazis.[135]

Despite all the signs that the National Socialists would soon take over, it was still a shock to Rudolphina when the Nazis annexed Austria and Hitler himself visited Linz on March 12, 1938. She noted that she "had loved Austria and her downfall hit me hard."[136] A feverish and frightening few months followed. Rudolph was briefly arrested again, the Menzels' passports were temporarily confiscated, their house was searched, and Rudolphina was publicly accused of being a communist despite her well-known antipathy to Bolshevism.[137] The city slaughterhouse was ordered to stop selling dog food to the Menzels, signs proclaiming "Jewish business" were plastered on their fence, and the neighborhood feud about the proper permitting for their kennel intensified. On May 25, 1938, the provincial government ordered the cessation of dog keeping on the Menzels' property and the "immediate removal of the Jewish kennel."[138] Rudolphina recalled that "we were now faced with the necessity to dismantle the entire basis of our life at short notice and with the severest material losses and accomplish our relocation quickly and contrary to our plans."[139]

The middle-aged Menzels rushed to pack up their home of sixteen years over the early summer months of 1938. Despite all their preparations for the move to Palestine and all the recent harassment they had endured, it was still difficult to abandon their home and leave their friends and colleagues, many of whom expressed deep dismay at the treatment they had suffered. The Menzels managed to get rid of most of their possessions, give away or sell most of their dogs, and send a shipment containing books, papers, linens, and typewriters to Palestine. Finally, on a hot August day, they said tearful goodbyes to their small staff and headed off to the train station in Linz with just two of their beloved Boxers; Rudolph held one leash and Rudolphina the other.[140]

The Menzels were traveling on forged documents arranged for them by a high-ranking official in the Hungarian Ministry of the Interior—an old friend from the cynological association. When they boarded the train, they were seated across from another well-dressed, middle-aged couple (the two Boxers were presumably sequestered in an adjacent compartment). Rudolphina described how the two couples glanced at each other in tense silence, alternately staring out the window and cautiously exam-

The Extraordinary Life of Rudolphina Menzel

ining each other as the train headed towards the Hungarian border. As soon as the wide Hungarian plains became visible, the man sitting across from Rudolph haltingly started a conversation in accented German, his tone hushed but angry as he decried the scourge of Nazism and the havoc it was wreaking on life in Vienna. His wife cried softly into her handkerchief as he spoke; Rudolphina noticed that stitched onto her sleeve was a small, defiant patch with the tricolors of the French flag. Her husband, whom Rudolphina described as an "Aryan" French businessman, had worked at a French firm in Vienna for many years and had spent the past months helping his employees flee after the Nazi annexation; now the time had come for him and his wife to escape themselves. As he described their plans to return to Paris, he became incensed about the French government's shameful acquiescence to the march of fascism across Europe.

The Menzels listened intently, trying to appear relaxed, but Rudolphina recounted how Rudolph nervously smoked cigarette after cigarette, knowing at any moment that they could be discovered, detained, searched, and arrested. As Jews traveling on illegal papers, they could not risk being openly critical of the Nazi regime with random strangers they encountered on a train. But as the Austrian border receded, and with it the possibility of being captured and imprisoned by the Gestapo, the Menzels evidently began to nod and smile at their French companions. Soon they were telling anecdotes and regaling the French couple with animated descriptions of their work and interests. Warm and vivacious conversation ensued until Rudolph reached in to his briefcase and removed a worn photograph of him and his wife posing with four handsome Boxer dogs in the leafy garden of their home in Kleinmunchen. He pulled out a fountain pen and wrote on the back in his rudimentary French, "Voyage a la liberté" and presented it to the other couple as a small souvenir of their brief encounter. The Frenchman in turn reached in to his pocket and pulled out a business card, on which he wrote in perfect German: "In memory of the day on which four people found their freedom again." When the couples departed the train in Budapest, Rudolphina described how they embraced warmly, "as if old friends" united by complex, shared feelings about the nightmare that was enveloping their beloved, free, and democratic Europe.[141]

Though Hungary was closed to Jewish refugees, the Menzels spent a few days outside of Budapest learning about best practices in sheepdog training, such as how to train dogs to bite without tearing the flesh of their flocks.[142] Rudolphina was apparently not going to pass up an opportunity to observe Hungarian shepherd dogs in action (which she considered the best herders in the world) just because they were fleeing for their lives and traveling illegally. After an otherwise arduous and tense journey, they arrived in Haifa in late fall 1938.

Rudolphina noted that she succumbed to only a single moment of nostalgia after she left Europe; she wrote to a kindly neighbor in Linz and asked them to go to her now abandoned yard, dig up some of her multicolored dahlia bulbs, and disinter the bones of her most beloved Boxers, which she had buried in a quiet corner of her garden. She politely requested the contents be placed in a box and sent to her in Palestine.[143] A neighbor later wrote that the Gestapo had come to pick up the Menzels three weeks after they left; it was unclear whether they were going to be detained because they were Jews or because Hitler sought Rudolphina's expertise at his military canine unit in Berlin. Their house was soon taken over by the Herman Goring Works, a Nazi industrial conglomerate.[144]

1938–1948: Creating the Canine Infrastructure in Palestine

> The people of Eretz Israel don't know all the pioneering work that went on behind the scenes in building our country. The hard labor that went into training military dogs was all done discreetly. To this day, the role of dogs in building our country has not been acknowledged; they were tools that built the country no less than the plow, the tractor, the gun and the water tower.
>
> RUDOLPHINA MENZEL[145]

After almost twenty years of British mandatory rule, Palestine in 1938 was a heterogeneous corner of a declining empire that was being unevenly policed by indifferent officials who were increasingly overwhelmed by intensifying local conflicts. The Menzels arrived in the country as part of a wave of immigration known as the "Fifth Aliyah," comprised of tens of thousands of well-educated, German-speaking Jewish refugees who emigrated to Palestine throughout the 1930s, forced to flee after Hitler assumed power. Amongst them were writers, scientists, and doctors as well as many of the best musicians in Europe. These new German-Jewish immigrants were viewed with a kind of *schadenfreude* by the Jews who had been building the country for the previous five decades, most of whom were from Eastern Europe. German Jews were accustomed to holding

The Extraordinary Life of Rudolphina Menzel 51

the upper hand in this well-known cultural divide. In Europe, cultured and cosmopolitan German Jews felt embarrassed by their impoverished, shtetl-dwelling brethren; in Palestine they were dependent on them for jobs, housing, and overall social welfare. The friction made for a lot of enmity, particularly since many of the German-Jewish immigrants expected they would be welcomed in Palestine as ambassadors of modern Western culture.

This Jewish culture gap was exacerbated by the fact that many of the more assimilated German-speaking Jewish refugees arrived in the country as devoted dog lovers. In some ways, this was good news for Rudolphina; she had a sympathetic, local audience who knew about and liked dogs and could help her with projects like organizing dog shows and properly breeding dogs. In other ways, the dog-loving habits of German-speaking Jewish refugees—particularly those who openly doted on their small, purebred companion dogs—only served to further alienate them from the local Jewish population. For the pioneering Zionist Jews in Palestine, caring for dogs was a waste of time, and feeding them was a particular affront at a time when food was scarce for people. To them, the German-Jewish penchant for pet dogs was a diasporic conceit, further proof that these new immigrants lacked any understanding of Zionist ideals and were unworthy of inclusion in their new Jewish society.

The Menzels were different. Unlike many of their German-Jewish brethren, they were lifelong Zionists, knew Hebrew, and had a very different understanding of and relationship to dogs. They also had the advantage of contacts made during their prior visits to the country and extensive professional connections, not only in the upper echelons of the Jewish Agency but amongst British Mandate officials as well.

When their boat docked in Haifa in the fall of 1938, the Menzels were met by Avraham Tzur, a graduate of one of Rudolphina's earlier dog-handling courses, who took them directly to his home and helped them get settled.[146] Shortly afterwards, the Jewish Agency found them a modest house in Kiryat Motzkin, a new suburb just outside of Haifa on the northern shore of Haifa Bay. Kiryat Motzkin had been founded in 1934, and by 1939, two thousand Jews lived in the town, which was composed of long lines of single-story, concrete houses covered in white plaster. The houses were spaced apart at regular intervals and many had front and back yards, just enough space to construct a few kennels and provide room for formal dog-training instruction. From the back door of their house on Barak Street, Rudolphina could look out and see far across the plains to the north; this would be her and Rudolph's home for the next thirty years.

After decades spent lecturing about Zionism, singing Zionist songs, and organizing Zionist activities from afar, the Menzels were about to start living its quotidian and challenging reality. This meant that, now well into middle age, they had to adjust to an entirely new way of life in which they lived in drastically reduced circumstances. Their house had only three rooms, rudimentary indoor plumbing, and intermittent electricity. In the summer it was unbearably hot, and in the winters surprisingly cold, the only heat supplied by an unreliable kerosene heater. They used one room as a bedroom and another as a living room, which was often occupied by one of the many live-in, teenage volunteers who soon staffed their growing kennel. The third room was reserved for Rudolph's office; he had minimal personal requirements but he did need one orderly space to work and write. This was a room Rudolphina was forbidden from entering lest it be overwhelmed by the whirlwind that always accompanied her.[147] Once their shipment from Kleinmunchen arrived, the Menzels had their typewriters, some of their books and papers, and a few basic pieces of furniture and bedding. There were small shops in Kiryat Motzkin that sold food, clothes, and other necessities, but the amenities were very basic, and they no longer had Rudolph's comfortable salary to purchase much at all.[148]

Undaunted, Rudolphina immediately affixed a sign to the outside of their house announcing the establishment of The Palestine Research Institute for Canine Psychology and Training and had official stationery printed with the institute's letterhead in Hebrew and English.[149] Modelled on the Austrian and German cynological institutes she knew so well, her institute would be the command and control center from which she would oversee all canine activities in the country. She would provide professional advice on veterinary health, regulate dog breeding, offer courses for dog trainers, manage dog shows, publish professional literature on dogs, set the criteria for dog licenses, and commence dog food production.[150] It would also serve as a base for her continued scientific research on the mental qualities of dogs and the evolutionary processes of domestication. From here, she would distribute her Hebrew-language dog training manual, which had since evolved into a lengthy textbook of over two hundred pages, featuring multiple chapters containing detailed instructions on how to train dogs to perform different tasks, such as how to climb ladders, pull ropes, rescue people on land or in water, and obey orders issued in whispers. It also included chapters on the dynamics of canine cognition, the principles of heredity, the impact of environmental influences on canine behavior, and the evolution of drives and instincts in the species *Canis familiaris*.[151]

The severe shortage of trainable dogs presented one of Rudolphina's most immediate and formidable challenges. Now that she was settled in the country, she could finally begin a more concerted effort to capture, test, and train the local variety of free-ranging dogs, some of which she knew would possess promising character traits. Catching and convincing semiwild dogs to remain with their human captors was not an easy task, however. Due to survival strategies evolved over millennia, most free-ranging dogs had never been handled by humans and were highly reactive, skittish, and suspicious. Moreover, as scavengers, these dogs were accustomed to guarding their resources, and most perceived any attempt to get close to them as a threat that must be aggressively fended off. Those that were more approachable were invariably covered with fleas and ticks, often suffered from mange or other skin diseases, and were frequently rabid.

Undeterred, Rudolphina and Rudolph set off to capture suitable specimens. She described how they staked out an observation point at the edge of a huge garbage dump on the outskirts of Tel Aviv and sat patiently "enveloped in clouds of evil odors" and "surrounded by swarms of flies and mosquitoes."[152] Using scraps of food as bait, they tried to lure the more curious dogs to come close enough that they could catch them; though most were resistant to the Menzels' entreaties, Rudolphina noted that they eventually managed to capture an injured juvenile dog and an abandoned litter of puppies.[153] After repeating this routine several times, she and Rudolph slowly gathered a small group of semiwild dogs in their Kiryat Motzkin yard, to which they added growing numbers of pedigree Boxers bred from the pair they brought with them as well as stray cats and even a few jackals.[154]

Rudolphina augmented the ranks of potential dogs for training by activating her preexisting canine networks. She asked prospective dog handlers in Jewish settlements to locate "young pariah puppies, about three to six months old" and to send her a written description of the puppies that looked most promising and, if possible, a photo.[155] Only after receiving Rudolphina's approval would a young dog then be invited for a temperament test, and only if it was found to possess desirable character traits would the dog and its handler gain admission to one of her free, ten-day guard-dog training courses. To attend these courses, handlers were asked to arrive in Kiryat Motzkin with their dogs, "bedding, a leash, a strong collar, and a dog bowl."[156]

By the end of 1939, Rudolphina reportedly conducted 235 temperament tests on a vast assortment of potential guard dogs (it is unclear how many were deemed suitable for training).[157] In the same year, her notes indicate that in addition to the dog-training courses she taught in Kiryat Motzkin, she spent 109 days on the road teaching dog-training courses at sixty-three settlements.[158] These exact numbers, however, may be slightly exaggerated; in a letter Rudolph wrote to Rudolphina's parents on January 30, 1940, he admitted that Rudolphina was very good at "sustaining the appearance" of activity so that they wouldn't be perceived as do-nothings (*batlanim*) or beggars (*schnorrers*).[159]

In a great testament to her cynological acumen, Rudolphina managed to achieve remarkable success in her efforts to train these "freshly-caught wild animals."[160] She reported that they made "astonishingly quick adjustments...towards tameness," and some were soon patrolling fields and guarding settlements; a bitch she named Hagar even became an outstanding mine-detecting dog.[161] Rudolphina noted that she and Rudolph were not the first to achieve success in training "wild" dogs; she credited the "Arab fellahin" with having long-established practices of using them "in primitive ways" to guard tents and flocks.[162]

Rudolphina's next task was to systematically breed the most outstanding and promising specimens of these dogs. Raising the subsequent puppies would be crucial to improving their quality for training since she could control their natal environments and socialize them during the critical early stages. But this proved not to be simple either since semiwild bitches have their own instinctual habits for nurturing their litters—none of which included human involvement—such as regurgitating their food, building underground dens, and fiercely defending any attempt to get close to their puppies. So Rudolphina took it upon herself to learn to negotiate their nurturing instincts and socialize their puppies. She then established stud books to keep track of their parentage and launched a formal breeding program for the free-ranging dogs in Palestine.

Rudolphina came to a series of optimistic conclusions regarding the potential of these dogs for service. Her optimism was informed by her understanding of the variability in canine temperaments regardless of breed but was also no doubt driven by the sense of practical urgency she

The Extraordinary Life of Rudolphina Menzel

felt in the face of such woefully limited local canine talent. They may have been stray, aggressive, fearful dogs, but because they were so distrustful and reactive, some could be successfully trained to be excellent, highly alert watch dogs. They may have been accustomed to living off of garbage dumps, but they had developed a concomitant acute sense of smell, which could make them good mine-detecting dogs. Most importantly, because they had learned to subsist on minimal food and water, tolerate intense desert heat, and survive desert sandstorms, they were perfectly adapted to the local conditions, unlike pedigreed imports. Rudolphina even came to the conclusion that these free-ranging dogs would make excellent pets; she managed to identify individuals who were not only loyal and inquisitive but who possessed the important character trait of "forgiveness": dogs that could be friendly towards humans even if they had been psychically or physically injured by them in the past.[163]

Collecting, testing, training, and breeding feral dogs was a necessary but not sufficient solution to the severe shortage of trainable dogs in Palestine. Rudolphina knew she was also going to have to create a framework for breeding dogs with known pedigrees and more predictable aptitudes. With help from the growing ranks of the Palestine Kennel Club, she launched an effort to record the origins and lineages of every pedigree dog in Palestine in breed-specific stud books. The stud books would serve as the foundation for regulating and organizing dog breeding in the country by establishing a clear, centralized basis for identifying, comparing, and selling dogs (although Rudolphina routinely made it clear that dog breeding in Palestine should never be undertaken for profit, only to encourage good practices).[164] Creating breed-specific stud books meant that dogs with known lineages and good temperaments could then be bought, sold, and mated together, creating a population of high-quality dogs that, if properly cared for, would be more valuable and trainable than random dogs of indeterminate provenance. Of course, the overwhelming majority of dogs in the area had no known pedigree, which meant Rudolphina had to start from scratch and do a lot of detective work.

Rudolphina appointed herself chief breed warden and as such was responsible for researching and managing all the breed-specific stud books in Palestine, a task she undertook with Teutonic attention to detail. Each dog's stud book registration not only recorded their gender, markings, and place of birth but it also included the dog's known family tree. More elaborate entries recorded relationships with dogs from all over the Middle East and as far away as India. These canine kinship charts revealed how purebred dogs regularly circulated between various British colonial territories. With the help of her fellow dog-loving compatriots, Rudolphina managed to record the family trees for 591 pedigree dogs in Palestine in 1939, and by October 31, 1945, her stud books contained entries for 2,083 dogs (Boxers were the most registered single breed by far with 1,371 entries).[165]

Rudolphina initiated these canine activities as World War II was raging in Europe and daily life in Palestine was growing ever more difficult. There were severe shortages of many household goods, food was rationed, and local tensions between Jews and Arabs were intensifying. Initial funds for Rudolphina's institute were not nearly what the Jewish Agency had promised, and in 1940, her budget was cut in half. In 1941, it was cut in half again.[166] Adding to these financial difficulties was the fact that the Jewish Agency's promise of a medical job for Rudolph long went unrealized.[167]

At home, Rudolphina often had to put her skills as a chemist to use in order to make soap; sometimes the Menzels even had to eat the discarded scraps of food they collected to feed their dogs.[168] One former kennel-helper who worked for the Menzels during these stressful years remembered once seeing Rudolphina kick one of her dogs when they did not obey her commands, an outburst entirely out of keeping with her generally indulgent attitude, particularly towards her Boxers.[169] It is unclear how Rudolphina found time between 1939 and 1940 to write her 261 page, single-spaced, typewritten memoir for the Harvard University essay contest "My Life in Germany Before and After January 30, 1933," but the $125 second place prize she received in February 1941 must have been very welcome ($125 is the equivalent of about $2500 today).[170]

These financial constraints, political exigencies, and personal frustrations did not prevent Rudolphina from spending the early 1940s creating a new national association for service dog trainers, consolidating the Dog Lovers Associations recently established in Tel Aviv, Haifa, and Jerusalem, and organizing dog shows, dog parades, and service dog performances. She and Rudolph always served as judges at these events, alongside an assortment of British Mandate officers and their wives.[171]

Rudolphina had long understood how a shared interest in dog sports could help the Jews establish good relations with the British.[172] She had curried favor with the wives of British police officers when she enlisted their support in the creation of the Palestine Kennel Club in 1937 and solidified patronage relationships with senior officials in the mandatory

The Extraordinary Life of Rudolphina Menzel

regime by honoring them with formal titles at local dog shows. By the early 1940s, Rudolphina had also developed numerous connections with British Mandate policemen, veterinarians, and public health officers; indeed, British colonial officials understood and appreciated Rudolphina's canine expertise probably more than anyone else in Palestine. In a report on a dog show in Haifa, Rudolphina made the benefits of these informal relationships explicit:

> This show was very successful, both in terms of the number of dogs shown and in terms of the many viewers who attended. The show carried much value, it must be stressed, not just in terms of examining the number of quality dogs that exist in the country, but also in terms of tightening the relations between our people and the rest of the nations, particularly the British people. 169 dogs participated in the show, 124 of them of service breeds.[173]

These connections ended up serving the British on a monumental scale during World War II, when they faced a desperate situation on the North African front as Allied forces became stymied by the Germans' massive land mine–laying campaign. Knowing that Rudolphina was a master trainer of mine-detecting dogs and a world-renowned authority on canine olfaction, the British asked for her help. After receiving official permission from Moshe Sharett—as well as a promise from the British never to use the dogs she gave them against Jews in the Yishuv—Rudolphina sprang into action. She issued a canine recruiting notice from her institute in Kiryat Motzkin seeking donations of healthy, sturdy dogs for war service and trained them using her incomparable methods.[174] She proceeded to supply the British army with four hundred expertly-trained mine-detecting dogs for deployment in North Africa.[175]

Closer to home, at the Palestine Kennel Club's 1942 dog show at the YMCA in Jerusalem, 129 dogs representing twenty-eight breeds were exhibited, including thirty-five Boxers, seventeen Cocker Spaniels, two Great Danes, and one Tibetan Terrier.[176] Dogs competed for a range of titles, awards, and diplomas, including Best in Show, Best in Active Service, and Best Young Dog Bred in Palestine. Shortly afterwards at a benefit dog show for the animal hospital at the old polo grounds on Bethlehem Road, prizes were given for the dog with the longest tail, the dog with the widest girth, and the dog with the most blood descendants present—categories clearly designed to amuse the audience and encourage attendees to keep track of their dogs' lineages.[177]

These events often attracted large crowds of curious onlookers and enabled Rudolphina to educate the public about both the genial nature of dogs in general and the remarkable abilities of service dogs in particular. Indeed, for those with little or no canine experience, dog shows provided a forum for people to observe, enjoy, and interact with dogs, perhaps for the first time. For those opposed to all things canine because they considered pet-keeping a bourgeois affectation, watching dogs climb ladders, scale walls, and crawl on command helped persuade them that dogs could actively contribute to the Zionist project by performing concrete tasks. As such, dog shows were a central part of Rudolphina's broader public relations campaign to encourage popular interest in and enthusiasm for dogs.

Dog shows served important social and political functions for dog breeders and trainers in Palestine as well. They provided a forum for them to exchange information, showcase their dogs' abilities, evaluate their dogs in comparison with others, and reinforce their sense of community. For those raising service dogs, the dog shows played an even more important role: they bolstered their communal feeling of being part of a national movement that was directly contributing to protecting the nascent state.[178]

In addition to these activities, Rudolphina routinely gave public lectures about dogs, wrote articles about dogs in local newspapers, and hosted events for reporters to watch dogs perform tricks.[179] She also began inviting curious neighborhood children to hold and play with her assorted litters of puppies in Kiryat Motzkin. This not only socialized the puppies but it also allowed the children to develop familiarity with and affection for dogs. Soon, regularly organized busloads of children from adjacent neighborhoods came to visit her growing menagerie, which included not just the progeny of her two Boxers from Kleinmunchen, but newly tamed stray dogs and other assorted, abandoned dogs as well. The Menzels' house in Kiryat Motzkin, where the "short fat woman with the dogs" lived, was soon a magnet for children all over the area.[180]

A contemporaneous article in the *Palestine Post* succinctly captured her proselytizing charisma during these years:

> Just as the Pied Piper of Hamelin attracted children with his music, Dr. Rudolfine [sic] Menzel attracts them with her knowledge of dogs. They have even coined a Hebrew title for the "calbanit" (dog woman)...when she appeared in the dining hall of Merhavia recently, a grapevine message seemed to shoot to all corners of the settlement and the children flocked around her...if Dr. Menzel has a way with dogs, she also has a rare charm with children.[181]

The Extraordinary Life of Rudolphina Menzel 59

Rudolphina's effort to familiarize Jews with dogs quickly gained traction in the broader Jewish culture. Friendly dogs first began to appear in Hebrew literature in 1943 when the popular writer Leah Goldberg described a woman with pet dogs, clearly modelled after Rudolphina, in her popular children's book *My Friends from Arnon Street*.[182] By the mid-1940s, a small network of dog-oriented services had emerged in Palestine. There were home kennels that bred and sold dogs, professional dog groomers, dog boarders, and private veterinarians. At a pet store called "All for the Dog" at 41 Trumpeldor Street in Tel Aviv, one could buy muzzles, harnesses, leashes, collars, and grooming supplies.[183] "Goliath Dog Biscuits" were made and sold in Petach Tikva.[184] "Dog medicines" to treat distemper, worms, and ticks were available for purchase at major pharmacies.[185]

In April 1944, the inaugural issue of *Hakalban*, Rudolphina's long-planned canine newsletter appeared. The lead article was by Martin Goldschmidt, the first president of the Palestine Kennel Club, and provides an illuminating glimpse into Palestinian canine culture at the time:

> If anyone had told me 20 years ago (when I acquired my first dog in the country) that someday there would be dog sports, exhibitions, shows, and even a dog journal here, I should not have believed it, any more than I should have thought possible many other things which have since happened or been achieved by us in Palestine. Who, in those days showed any interest in dogs except for some Englishmen and a few "crazy" Jews! Systematic breeding was not practiced at all, except for the existence of a very small number of shepherd dog strains, and the dogs which were trained in this country either for a practical purpose, that is to help their masters, to work with their masters, or out of a sporting interest, could be counted on the fingers of the hands. Nowadays, there are whole generations of dogs reared by the most up-to-date method, a stud book for the registration, courses for the instructions of both dogs and leaders, exhibitions, and a dog loving community which takes an eager interest in everything connected with canine affairs.[186]

This mimeographed, type-written publication came out almost monthly until 1947 and was sold in Palestine for a few piasters.[187] Subsequent issues featured announcements about dog-training courses, dog shows, and dog parades as well as minutes from the monthly Dog Lovers Association meetings.[188] They also contained local and international canine news, frequent reports on the successes of war dogs in Europe, and special features on issues like the legal liabilities for dog bites, instructions on how to breed a bitch in heat, and how to whelp puppies during a *khamsin* (the oppressively hot desert wind storms that periodically swept through Palestine).[189]

Breathless accounts of local canine heroics also appeared frequently in the pages of *Hakalban*. For example, in the September/October 1944 issue, Paul Marx of Kibbutz Hazorea writes:[190]

> I wish to report on the work of my dog "Ted" thanks to whom I was able to arrest a thief in the orchards of our settlement and thus capture a long-wanted gang of thieves. Ted is about three years old and has spent almost his whole life with me. He is a cross between a Boxer and a Great Dane. He has had no thorough training, never attended any regular training course, but only took part in a few training days. Recently he again accompanied me on night duty in our orchards. When the thieves approached, he at once obeyed my order to go for them, seizing one of them and throwing him to the ground. The other four were panic-stricken and fled. Ted remained sitting at the head of the prostrate man, guarding him excitedly but without further attacking him. When his accomplices returned to rescue him and recover the loot, he charged them again, upon my order, and chased them away. They left everything behind, their sacks and their donkeys, and made off. The dog returned to me and did not leave me again. The captured thief, in his terror of the dog, gave away his friends and now five offenders are facing the judge.[191]

This sort of personal testimony illustrates how valuable guard and patrol dogs had become in Palestine—in fact, by the mid-1940s, most Jewish settlements were protected by dogs.

Many of Rudolphina's articles in *Hakalban* described the rigorous regimens of specialized care that must be provided to keep dogs healthy, especially in hot desert climates; these articles illustrate the enormous amount of time and expertise that was required to effectively care for dogs, particularly imported pedigree dogs that were not adapted to the local conditions. Dog owners were advised to conduct daily inspections for ticks and other disease-carrying vermin by carefully combing through their dogs' fur with a special steel comb and were admonished to bathe their dogs thoroughly at least twice a week. Those caring for longhaired dogs were advised to keep their coats clipped short, but not too short, so as to not overexpose their canine skin to the desert sun. The heat posed such a considerable danger to dogs that Rudolphina routinely proffered concrete advice about how to keep dogs cool: for example, by ensuring that clean water was always available, exercising dogs only in the cooler parts of the day, and wetting their fur to cool them down.[192] Sometimes she would include meticulous physiological explanations of the processes

The Extraordinary Life of Rudolphina Menzel

of canine perspiration for no other purpose than to satisfy the scientific interest of her readers.[193]

Canine cleanliness was so important that Rudolphina provided specific instruction on how to build and disinfect kennels:

> It is impossible to keep a dog clean without maintaining cleanliness in his dwelling or sleeping place... in our climate, the very construction of the kennel should be different. It should, if possible, not consist of wood but of brick or concrete, and have a concrete stone floor; it can then be cleaned just like a human dwelling. It goes without saying that the kennel must be cleaned thoroughly every day. In the rainy season it is sufficient to wipe off the excrements with sand, saw-dust, ashes, or the like, to sweep thoroughly afterwards and to rinse every other day with a disinfectant solution. A solution of creolin (3 tablespoons in 1 liter of water) or a hot solution of soda is a suitable disinfectant. In the hot season, this kind of rinsing must be done daily, and it is absolutely imperative that food remnants and excrements should be removed *instantly* [emphasis in original]. It does not do to leave them about until the time of the daily cleaning, as this would attract swarms of flies... Special stress must be laid on the cleanliness of the sleeping area, which should be cleaned thoroughly with a soap and brushed with disinfectant solution, particularly in the corners and crevices, where ticks like to sit and deposit their eggs... it advisable to spray the corners and crevices with a strong solution of flit prior to washing, this will force the parasites to come out... *No dog, whether he is kept in the house or in a kennel, can be healthy and serviceable without cleanliness. In our climate, more than in any other, a person who has no time or chance to keep his dog clean should not have a dog at all* [emphasis in original].[194]

In an article on the dangers of allowing dogs to roam freely, Rudolphina provides an additional glimpse of the general challenges facing dog owners at the time.

> Many people think that in order to give their dogs sufficient exercise, they must leave them to stray, if they themselves have no time to go with them. We are definite opponents of straying, also in colder climates. It corrupts to [sic] the dog, makes him practically unfit for the watch and guard service, and involves numerous dangers for the dog. In our climate, the dangers are multiplied. Jackals and other wild animals do not submit to any rabies order, and the many wild and half-wild cats and dogs can hardly be controlled by the Government veterinary. To this is added the danger of infection with other diseases, especially many skin diseases (scabies) and various worms, the danger of eating carrion or poisonous baits, (and) the drinking of contaminated water.[195]

Her grave concerns about the considerable dangers of "straying" remind us that most dogs in 1940s Palestine were still ownerless stray dogs that foraged and marauded around villages, towns, and cities, often while infected with lethal diseases. This was a tricky context for Rudolphina to impose her plans for a well-regulated canine infrastructure. Not only did she have to overcome the traditional Jewish ambivalence towards dogs as well as the Zionist resistance to them as bourgeois affectations, she had to do so while negotiating the legitimate widespread fear that dogs were unpredictable, dangerous, and rabid.

Occasionally, Rudolphina also penned scholarly articles for *Hakalban* on topics like the "Genetic Components of the Canine Character" or "The Dog's Visual Faculty"; the latter included results from recent experiments she had conducted on the ability of dogs to recognize their owners and at what distances.[196] She frequently wrote articles reminding readers about the importance of canine temperament testing and its role in improving the quality and performance of service dogs.[197]

Obituaries for both notable dogs and their caretakers also appeared. The September 1944 issue included a regretful announcement about the death of "a very good Alsatian named Anak...who spent many years of his life in the security service. His good character he transmitted to his children. One of his sons serves with the Military Police in the Middle East. Our sympathy goes out to his owner, Mr. Yitzhak Neuberger of Nahariyah."[198] The reference to Anak's son clearly suggests that dogs bred by Jews in Palestine were being sold or given to British officers for use in the larger mandatory regime.

One of the more poignant human obituaries, written by Rudolphina herself, was a tribute to Paul Bonem, a Jerusalem-based, German-born gynecologist whom she later identified as a Haganah arms smuggler.[199] Bonem passed away at the age of fifty-seven and Rudolphina fondly recalled the last time she saw her friend at an advanced dog-training course for service dogs:

> On the last course day...towards midnight we were marching home from night practice...by the light of the moon we passed through somnolent villages, singing in subdued tones. We were all tired, very, very tired, but in excellent spirits; and Paul Bonem kept pace and step with the youngest of us. Afterwards, we sat together a long time on the terrace, he pointed out how within the brief period of the course a motley crown of dog-trainers from "Dan to Beersheba" had been welded into a community and he said how pleased he was with the course and how much it reminded him of the good old days at Mikve.[200]

The Extraordinary Life of Rudolphina Menzel 63

This short excerpt not only provides a rare mention of the clandestine, nighttime dog-training courses Rudolphina was conducting at the time, it offers insight into the appeal of dog sports for German-speaking Jews, particularly those like Rudolphina who had spent their teenage years hiking and singing with their *Blau-Weiss* comrades in Central Europe before World War I. To honor Dr. Bonem's memory, members of the Dog Lovers Association created the Dr. Paul Bonem Fund to support "all dog purposes in the country."

In September 1944, over a hundred people gathered at the popular Café Casino on the beach in Tel Aviv to hear Rudolphina give a lecture on "the importance of dogs for the national home." (According to Rudolphina's own report on the event, "her splendid explanations were awarded by stormy applause and cheers").[201] In November 1944, local screenings of the movie *Lassie Come Home* at a Tel Aviv cinema provided material for lengthy debates amongst all the Dog Lovers Associations about whether a lost dog could indeed find its way home.[202] The film was declared "most effective propaganda for our aims amongst the public," presumably because it would cause audience members to fall in love with Lassie and make them more sympathetic to dogs in general.[203]

On April 15, 1945, the Dog Lovers Associations organized their largest event up to that point: a national Palestinian dog show at Maccabi Stadium in Tel Aviv, billed as "the first Service-dog performance in our Jewish City."[204] Buses were chartered to bring participants to the event from Haifa, Jerusalem, the Jezreel Valley, and the Shomron area on the slopes of the West Bank.[205] Though there were apparently snafus with the loudspeaker system, numerous delays due to canine recalcitrance, misunderstandings about the necessity of washing dogs before they were shown, and vocal complaints that Rudolphina's favorite breed (Boxers) were overrepresented, eleven prizes were awarded.[206] A wire-haired terrier named Ronny from Haifa won Best in Show, a Boxer named Gid from Holon won Best Service Dog, and Best Puppy was awarded to a Scotch Terrier named Babette from Mizpah.[207] In addition, twenty-six dogs received marks of distinction, including Japha, an Alsatian from Tel Aviv; Lola, a Boxer from Ramat Gan; and Sepell, a dachshund from Haifa.[208] That the event was held under the "distinguished joint patronage of R. H. R. Church, ESQ. and District Commissioner and Brigadier C. Greenslate CBE Area Commander in aid of the British Red Cross, the SPCA Tel Aviv-Jaffa, and the Jewish Soldiers Welfare Committee" provides another vivid illustration of the way canine sports had come to informally ingratiate dog-loving Jews with the upper echelons of the mandatory regime.

In the April 1946 "Purim" edition of *Halkalban*, Rudolphina published a whimsical "Dog Ten Commandments."[209] These "commandments" are not just amusing; by reading their inversion, they illustrate the common attitudes towards dog owners at the time. Commandment number nine offers a particularly humorous commentary on popular perceptions of German-speaking Jews and their companion dogs as well as Rudolphina's understanding of the cultural danger this association represented for the larger canine cause.

1. Your dog is the center for the world, nothing equals him in importance.
2. When you take him out in the traffic, under no circumstances put him on a lead, not to prevent him from running between passers-by legs, to dirty their clothes, to throw over little children, or at least to frighten them, to chase cyclists and so on. The same applies to public places, shops, restaurants, buses, etc.
3. When he wants to deposit his feces in the middle of the pavement or "sign" food stands with his urine, don't hinder him, he is quite right.
4. At the butcher or grocer tell detailed stories about how spoiled your dog is and what all he does not eat. Buy a piece of steak for him that makes a good impression, especially if the shop is packed with people.
5. Only your dog is of importance not the dog of your fellow-creatures. In case of a fight, the other dog is always wrong.
6. At home let him bark and howl as much as he likes. The neighbors have nothing to do with it as it is *your* dog.
7. Allow him to bite the postman, the milkman, etc. If he likes it and you enjoy it, it is quite in order.
8. If you don't need to keep him clean from flies and ticks, that would mean to drive things too far, as he is after all only a dog.
9. In public speak to him much and as loud as possible in German, that makes many new friends for the dog-cause.
10. Try your best wherever you can, to molest and anger other people with your dog, as that is why you possess him.

Much of the information and advice presented in *Hakalban*, particularly pertaining to canine care in desert climates, was of great interest to dog lovers not just in Palestine but in other countries as well. Indeed, Rudolphina sent copies of the English language version of *Hakalban* called *The Palestine Kennel Gazette* to dog lovers in France, Poland, South Africa, Canada, and the United States as well as Uganda, Turkey, and neighboring Arab countries.[210] In this way, her newsletter served

The Extraordinary Life of Rudolphina Menzel 65

as a form of canine diplomacy that created pathways for all kinds of informal, international connections.[211] While it's unclear exactly what concrete purpose these connections served, they certainly familiarized people around the world with dog-related activities in Palestine and by extension perhaps created sympathy towards, and a sense of allegiance with, Jewish dog-owners there.

After World War II ended, the British became increasingly hostile to the local Jewish population and tensions with the local Arab populations intensified. As it became clear that the Mandate was coming to an end and that the British would leave the region for good, the leaders of the Yishuv began preparing for a broader conflict with the local Arabs. They started drawing up battle plans, collecting weaponry, and stockpiling arms.[212] They knew that any war with the Arabs would be fought at close range and in rocky, hilly terrain. Most battles would be waged between small groups of combatants using hit and run tactics like raids, ambushes, and sabotage. In such conditions, when communication between soldiers is essential but dangerous and supply lines are attenuated and improvised, using dogs could provide both strategic benefits and psychological advantages. Canine soldiers could not only execute the tasks they had been trained to perform, they could effectively frighten, intimidate and subdue their adversaries, thereby providing their handlers with an additional layer of safety and protection.[213]

Moshe Dayan, then a senior commander in the Haganah, had already appealed to Rudolphina to start teaching dog-training courses that focused less on defense and protection and more on utilizing dogs as weapons of war. In response, Rudolphina spent the mid-1940s building up not only a cadre of highly trained and tested dog-handling experts but a unit of highly trained and effective canine soldiers as well. To avoid British suspicion, she used her Dog Lovers Associations as a cover for these military dog-training activities.[214]

Rudolphina trained messenger and ammunition dogs to run between a fixed point and a moving point so that messages and ammunition could be carried from command centers directly to soldiers in the field, even under conditions of heavy gunfire. Dogs were trained to cover distances of up to a kilometer in as little as two minutes; if scent-trails were laid between two points, dogs could reliably expand their range up to five kilometers.[215] Messages were affixed to dogs' collars, and specially made canine saddles were attached to their backs, enabling them to transport up to six kilograms of ammunition or medical supplies, a significant amount for soldiers fighting or wounded in remote areas.[216] This was difficult work in the local topography, where shifting desert winds and sandy soil made scent-trails hard to follow, thorny underbrush and cactus outcroppings stymied canine pursuit, and feral dogs and cats, let alone wild jackals, provided ever-present distractions.

Laying mines had been a frequent tactic of Arab resistance in the years leading up to the 1948 war, and Rudolphina had already established a unit of mine-detecting dogs and handlers, but as preparations for a regional conflict accelerated, she trained even more. These dogs were trained to walk cautiously in front of their handlers on forty-foot leads and to signal their handlers when they detected any kind of disturbance in the dirt or sand. The dog had to be trained to detect the smell of the lesion in the ground, not the smell of the metal in the mine—if the mine was buried too close to the surface, the dog would explode.

Rudolphina also trained dogs for kamikaze missions to destroy enemy tanks using a tactic she borrowed from the Russians, who used "anti-tank" dogs extensively in World War II.[217] These dogs were relatively easy to train: a hungry dog was repeatedly lured under a truck with the reward of food. Soon, the dog learned to associate food with the undercarriage of a large rumbling vehicle. In wartime, a bomb could be strapped to the hungry dog, which would then explode when dispatched to the enemy tank, blowing up both the tank and the dog. For this grim assignment, Rudolphina trained dogs with relatively mediocre aptitudes so she didn't have to sacrifice those with more valuable temperaments.

Training patrol dogs was also imperative for preparing for the coming conflict. These dogs were taught how to alert their human companions to hidden ambushes by walking in front of soldiers while tethered on long leads before lunging and biting on command, making them dangerous and effective weapons in face-to-face combat.[218]

After British police dogs uncovered a Haganah stockpile of ammunition in 1946, Rudolphina also began teaching underground courses in canine olfaction.[219] This enabled Haganah operatives to learn how to disguise the smell of gunpowder, create decoy ammunition stashes, and construct hiding places without leaving mounds of dirt since the smell of loose dirt was easily detectable by tracking dogs.[220] By confusing and confounding British scent hounds, these olfactory ruses played a pivotal role in the years leading up to the 1948 war, when ammunition was scarce and protecting it provided an important tactical advantage.[221]

The Extraordinary Life of Rudolphina Menzel 67

As tensions mounted between Arabs and Jews in the final months of 1947, Rudolphina was put in charge of directing a more formal recruiting effort to enlist "four-legged soldiers." She was driven all over the country in a pickup truck filled with wooden crates in order to find and collect suitable dogs to press into service. Some were pet dogs donated by their owners, while others were stray dogs and puppies that had been rounded up and brought to established locations. At each stop, Rudolphina hopped out of the pickup truck and conducted her temperament tests in order to evaluate and sort the dogs for relevant training. Those with promising character traits were put in crates, loaded on the pickup, and transported either to Kiryat Motzkin or to one of the two newly established satellite dog-training stations: one in north Tel Aviv and one in the Valley of the Cross in Jerusalem.[222]

The civil war between Palestinian Jews and Arabs that erupted in response to the United Nations Partition Plan in November 1947 was the culmination of hostility and antagonism that had been building for decades. The ensuing conflagration was joined by armies from the surrounding Arab states, and battles raged between the two sides throughout the winter. It wasn't until March 1, 1948 that the official canine unit of the Haganah finally opened in an abandoned British military outpost in Kiryat Haim, the town adjacent to Kiryat Motzkin.[223] The unit was part of a larger Haganah animal corps in which mules, camels, and donkeys were trained to transport supplies, and pigeons were used to carry messages between a large, clandestine network of dovecotes throughout the country.[224] Levi Eshkol, Israel's first director of the Ministry of Defense (and later Israel's third Prime Minister) reportedly joked that if food stores ran low, the pigeons could be fed to the dogs.[225]

Rudolphina naturally assumed that as the person who had created and directed the Jewish militia's entire military dog-training program and who was the supreme canine authority in the country, she would be appointed commander of the new canine unit. But Rudolphina's age, gender, and perhaps even her Austrian-ness finally caught up with her. Now that they were at war, the young Jewish soldiers weren't going to take military orders from an eccentric, fifty-seven-year-old woman with a heavy German accent, no matter how much of an expert she was.

Rudolphina had effectively been sidelined by some of her own protégés. Against extraordinary odds, she had managed to establish a substantial corps of experienced dog handlers, many of whom had long been chafing under the authoritarian control she maintained over every dog-related activity in the country.[226] When they saw an opportunity to finally challenge her authority, they did. Though Rudolphina was not summarily dismissed or publicly humiliated, she was demoted to a part-time educational supervisor, and her former student Abraham Zirlin assumed command of the new canine corps.

The first soldiers to serve in the unit were graduates of Rudolphina's earlier dog-training courses—two of whom, Alexander Cohn and his wife Ester, later wrote down their recollections of this time. They remembered the unit as initially very disorganized, and noted that no one really knew what they were doing.[227] Their exhausting days began at 4 a.m. with intensive exercises to prepare dogs for combat. These included drills to condition the dogs to loud noises and to increase canine endurance by training the dogs to run alongside their bicycles as they pedaled as fast as they could. In her role as educational supervisor, Rudolphina gave lectures on canine behavior to the new recruits and taught them how to train dogs to search for mines, detect ambushes, and carry messages. The soldiers were required to keep detailed records of each dog's successes and failures when performing particular tasks and were subject to rigorous exams administered by Rudolphina herself. As the war went on, new recruits were randomly assigned to the unit, many of them Holocaust survivors.

Ester Cohn later estimated that at least three hundred dogs were operationalized during the conflict between 1947 and 1949, and while there appear to be no records of specific canine achievements during the war, there was certainly widespread understanding that dogs had contributed to the Jewish victory. Columns of military service dogs and their handlers proudly marched in the military victory parades during the summer of 1949 after the armistice agreements were signed; by that time thousands had been killed on both sides, and almost one million Palestinians were displaced; over 700,000 became refugees.

Of the many asymmetries in the first Arab-Israeli war, this one has been traditionally overlooked by historians: dogs were deployed only by the Jewish forces. Palestinian Arabs were certainly no strangers to military dogs; they had been contending with British Mandate police dogs for almost fifteen years before the 1948 war broke out. In fact, stories about their fearsome capabilities and their capacities to track criminals and terrorize suspects were well-known, even in the most remote Arab villages.[228] But traditional Islamic beliefs about the uncleanliness of dogs, let alone the complete absence of any kind of infrastructure for systematically breeding and training them, meant Arab fighters could only try to avoid, outwit, poison, shoot, or stab enemy dogs rather than enlist them to fight on their side.[229]

1949–1973: Guide Dogs and Canaan Dogs

Despite Rudolphina's efforts to save it, the Research Institute for Canine Psychology and Training in Kiryat Motzkin was officially dissolved in 1949, and both its functions and its dogs were absorbed into the new Israeli army and police canine units. By this point, the Menzels were in their late fifties, had been in Palestine for over ten tumultuous years, and had lived through unprecedented personal, political, and cultural upheavals even before they arrived in the country. Rudolphina had achieved some extraordinary successes but had also encountered unending professional obstacles, financial difficulties, and personal challenges. While she had cultivated a devoted circle of professional colleagues, she had also managed to ruffle a lot of feathers.[230] Many of her students and colleagues found her brilliance and charisma irresistible; others found her domineering, short-tempered, and sometimes obtuse.[231]

In addition, the Menzels were now living with the knowledge that the Austria of their youth had been decimated in World War II. Though most of their extended family had managed to escape to America before the Holocaust, their growing understanding about what the Nazis had done to the Jews of Europe was simply overwhelming. It wasn't until almost two decades later that Rudolphina reflected publicly about the fact that dogs descended from her beloved Mowgli, and trained using methods she had developed, were subsequently operationalized by the Nazis to chase, terrorize, and kill Jews and scores of others throughout the reign of the Third Reich: "I suffered a lot knowing that my students in Austria and Germany were using the knowledge they acquired from me to use dogs to exterminate my people and other peoples."[232]

Rudolphina took stock of her situation. She was fifty-eight years old and could very well have chosen to slow down, rest, or even retire. Instead, she decided it would be a good time to pivot and launch an entirely new professional career. It so happened that Israel in the late 1940s was home to a growing and diverse population of blind people, a disproportionate percentage of whom were of working age. Some were blind from birth or childhood, some had lost their sight in the 1948 war, others in the wreckage of postwar Europe, and still others were amongst the ever-increasing number of Jewish immigrants from Middle Eastern countries who were sightless when they arrived in Israel. Rudolphina recognized an opportunity and decided to create the Israel Institute for Orientation and Mobility of the Blind, the first school for guide dogs in the Middle East.

Rudolphina was familiar with guide-dog training from her cynological work in Austria.[233] She had developed temperament tests for guide dogs in her foundational work on the subject, and the basic principles of guide-dog training drew directly on the same kinds of scientific questions and practical conundrums that had long fascinated her. But even with her extensive expertise, guide-dog training presented her with a new and different set of challenges.

Unlike other kinds of service dogs that are trained to directly obey the commands of their handler (in and of itself an often complicated task), the guide dog must be taught how to strategically disobey their handler. For example, if a blind person commands their dog to walk into traffic, the dog must know not to follow that command. Teaching a dog to discern between what their handlers tell them to do and what they need to do to protect their handlers requires acute canine intelligence and astute human understanding of canine cognition. The underlying concept of "sensible disobedience" presents a complicated conundrum for the dog as well as a very compelling training problem for the handler. Rudolphina happily immersed herself in this new set of puzzles and soon mastered all the relevant practical and theoretical literature on guide-dog training.

Launching this new endeavor mobilized Rudolphina's socialist impulses as well. As she learned more about guide dogs and the ways they increase mobility and independence for the blind, she recognized how crucial they were for enabling blind people to find work in mainstream industries, not only in the traditional "blind man's trades" of chair caning, street vending, and switchboard operating. She came to believe that guide dogs promised to make the world a fairer place by empowering blind workers to retain economic equality with sighted people.[234] Guide dogs also provided additional benefits: they gave the blind greater personal autonomy by relieving relatives from the responsibility of serving as escorts. For blind people who lived alone, guide dogs could also provide welcome companionship.[235]

Once she had mastered the new theories and methods on which guide-dog training was based and identified the problems that guide dogs could

The Extraordinary Life of Rudolphina Menzel 71

solve, Rudolphina turned to the pragmatic, cultural, and logistical challenges her new institute faced. The Ministry of Social Welfare and the Veterans Department of the new Israeli army helped her secure startup funds and a small campus for her institute in an old British army barracks in nearby Kiryat Haim. Volunteers built an obstacle course on the property to serve as a preliminary training area (one of the foundational skills for a guide dog is to learn how to navigate around obstacles). Then came the familiar challenge of raising money, both to teach and pay dog handlers and to buy and care for dogs.

Next, there was the task of recruiting expert handlers and assembling a robust group of suitable dogs to train. Recruiting handlers was relatively easy. Rudolphina had maintained a strong network of former students who were eager to continue working with her, but finding suitable dogs for guide-dog training posed more of a problem. Guide dogs must possess particularly outstanding temperaments; they cannot be aggressive or easily distractable, and they must be exceptionally cooperative and loyal. The canine corps of the new Israeli army had absorbed the finest of her trained military dogs. All of the excellent Doberman Pinschers in the canine unit of the Palestine police had been euthanized by the British as soon as the Mandate ended. Pedigree dog breeding in the country was still woefully limited, and those dogs that were selectively bred were often spaniels, hounds, or terriers—breeds that were usually too easily sidetracked by odors to serve as reliable guide dogs. So Rudolphina started again from scratch, recruiting all manner of dogs from her canine networks and conducting temperament tests on all the potential draftees. She also wrote to her former cynological colleagues in Austria asking them to send her eighty German Shepherds between the ages of nine and eighteen months old—appearance irrelevant, but outstanding disposition essential.[236] By mobilizing her networks and conducting dozens of temperament tests, she gradually managed to collect and train an unlikely assortment of potential guide dogs.

Finally, Rudolphina had to recruit blind students willing to learn how to depend on and live with guide dogs, which of course meant they couldn't fear or despise them. This was a cultural challenge she was very familiar with, but it was compounded by new socioeconomic complexities facing the thousands of immigrants who arrived in the country after the establishment of the state. By 1956, eighty-seven percent of the blind population in Israel consisted of recent Jewish immigrants from Middle

Eastern countries.[237] Like their brethren from the ghettoes of Eastern Europe, these Jews were coming from communities that traditionally feared or despised dogs; before they would even consider using guide dogs, they would have to learn to tolerate dogs in general. Even if they could overcome their traditional fear of dogs and learn to depend on guide dogs for increased mobility outside their homes, they often lived in small apartments with extended families, surrounded by relatives who were not predisposed to living with dogs, regardless of their utility. In addition, many of them didn't yet speak modern Hebrew, which would make it difficult to work and live with a guide dog trained to respond to Hebrew commands.

Rudolphina knew the program's success depended on pairing guide dogs with highly motivated blind handlers, and since each guide dog represented such an enormous investment of time, money, and expertise, carefully vetting prospective students was critical. So Rudolphina developed an elaborate application process, including detailed interviews in which applicants were asked about their personal background, living arrangements, overall medical condition, and employment status. It was very important to Rudolphina that both Arab and Jewish applicants were considered equally, and she made sure to exclude IQ as a prerequisite for admission since she saw no correlation between intelligence and success with mobility training. Those who managed to be admitted would be paired with a dog and readied to begin their lengthy residential training program. They would spend twenty-four hours a day with their dogs and would not only learn how to train and work with their new canine companions but they would learn how to feed, groom, and care for them as well.

It is important to note that Rudolphina was not the first to recognize the needs of the blind in Palestine. Since 1924, the Palestine Lighthouse had been running a school in Jerusalem that offered instruction in Hebrew braille and vocational programs for both residential and day students (after World War II, the organization worked with the Joint Distribution Committee to rescue blind Jewish refugees from Europe and bring them to Palestine).[238] In fact, through connections Rudolphina established with the Lighthouse, Rudolphina persuaded Helen Keller, who was on their board, to come to Israel and give the inaugural address at the long-awaited opening of her Israel Institute for Orientation and Mobility of the Blind in 1952. Helen Keller began her address thus:

Friends, it is an honor to me to salute you, workers and creators of a new commonwealth in a land echoing with history through the ages and associated with the Bible, my best loved book. Since you are in the vanguard of progress aiding all classes of the handicapped, I am happy to talk to you about the blind whom I represent... Confidently I look to you, dear friends, to do unto my blind fellows in Palestine as you would have others do unto you. If you do, the blessings of sight you enjoy will be all the sweeter to you.[259]

Rudolphina's relationship with the Lighthouse ultimately soured, for reasons that remain unclear, which meant that she could not rely on their established network of devoted American donors. As a result, she had to spend considerable time and energy fundraising herself. The institute came to serve the needs of an exceptionally diverse population of blind people from different countries and backgrounds, whose blindness had various origins, congenital or adventitious. In addition to teaching students how to live and work with guide dogs, the institute gradually evolved into a social welfare organization that helped its students learn Hebrew, find suitable housing, and work and adapt to their new surroundings, all during a time of general austerity in Israel.

As the institute's ambit expanded throughout the 1950s, Rudolphina became increasingly committed to understanding the complex dynamics that shaped the needs and experiences of her sightless students. In addition to analyzing various logistical, psychological, and cultural factors that shaped a person's attitude toward their disability, she was particularly interested in exploring how the origin and etiology of a person's blindness affected their motivation and capacity for different modes of mobility training. In order to investigate these and other sensitive questions, Rudolphina assembled a research team and began applying for grants as Principal Investigator. This necessitated her to acquire a new range of theoretical expertise, master a new set of qualitative tools, and familiarize herself with a new international community of professional researchers. As with her first cynological projects, Rudolphina had both a practical agenda—to improve the lives of blind people—and a scientific agenda—to better understand the cognitive faculty of perception.

In 1963, Rudolphina organized a visit to Israel for a delegation of American mobility experts from the United States Department of Health, Education, and Welfare. The following year, she was awarded a multi-year research grant from the US government to explore the different capacities for orientation and mobility of those who were born blind and those who were blinded later in life.[240]

To collect data, Rudolphina and her team developed a series of extensive questionnaires that they administered to hundreds of blind Israelis of various sexes, class backgrounds, ages, and places of origin. They also conducted 367 in-person interviews with blind Israelis, their families, neighbors, and friends. The final 117-page report "The Importance of Orientation and Mobility for the Blind" was issued in English in 1969.[241] It contained dozens of charts and graphs that illustrated how increased mobility—whether provided by the use of a guide dog, a cane, or both—affected the economic prospects, social experiences, family relationships, and levels of self-confidence of those interviewed. It also included personal testimonies such as this one from a Jewish woman in Haifa: "The day I began traveling, guided by my dog, was an experience I shall not forget... I was free all at once from all strain and tension... Suddenly I sense a special satisfaction while walking—a very distinctive feeling which I had already forgotten." Another Jewish woman, identified as a mother of two children who was losing her sight due to glaucoma, is quoted as saying "mobility was the cornerstone of my rehabilitation." An unnamed blind Arab man reported: "Three years have passed since I was made an independent man by receiving 'Lady' my guide dog from your school. Those three years were perhaps the most important years of my life, and certainly the most useful. Had it not been for my dog, I do not think I would have been able to achieve what I desired, at least not in such a short time."[242]

Perhaps most importantly, the final report contained an updated version of Rudolphina's "mobility-training readiness test," which she first published in 1967.[243] This detailed eleven-page questionnaire was designed to elicit specific information about the particular mobility needs of a blind person depending on their individual disposition and personal circumstances. Rudolphina's mobility readiness test, the first of its kind, filled an "urgent need" to identify the largest number of blind people who could benefit from mobility training and was distributed to a range of relevant organizations overseas.[244] In sum, throughout the 1950s and '60s, Rudolphina remade herself yet again, this time as an energetic and influential figure in the international community of researchers and practitioners devoted to better understanding and attending to the mobility needs of the blind.

The Institute for Orientation and Mobility was not Rudolphina's only ambitious new undertaking in the last decades of her life. She also decided

The Extraordinary Life of Rudolphina Menzel 75

to return to her scientific research on the local, free-ranging dogs, which had fascinated her since her first visit to the country in 1934. Though she had already spent years capturing, training, and breeding select specimens in order to alleviate the shortage of trainable dogs in the country, her pragmatic interest in these dogs had always remained secondary to her scientific interest.

Her research was based on a theoretical assumption she shared with her cynological mentor Emil Hauck: that these semiwild dogs represented a distinct, intermediary breed that had somehow been frozen in type and unchanged in temperament since Neolithic man first tamed wolves and made them into dogs. As such, they could turn out to be the missing link between wolves and dogs. If Rudolphina could identify, isolate, and preserve "pariah" dogs that constituted an ascending evolutionary development from the wolf to the domestic dog (rather than a descending development or an aggregate, genetic admixture of all domestic dogs), her results would be of great interest not just to cynologists, but to all scientists interested in the evolutionary processes of domestication.[245]

Rudolphina had already spent years gathering data on how free-ranging dogs learned to acclimate and adapt to their new surroundings depending on differences in their ages and individual traits.[246] To investigate the evolutionary steps between wildness, tameness, domestication, and trainability, she used this data to discern how and when individual behaviors gradually changed from skittish and unpredictable to tame and obedient. She recorded how and when being fed, washed, and cared for changed a dog's affect from wariness to friendliness. She tried to understand whether their new behaviors were permanent and predictable or easily forgotten. Further, she studied whether a tamed dog's newfound friendliness extended to all humans or only to those recognized as new members of the dog's "pack."

Rudolphina's Zionism had coincidentally enabled her to become the first cynologist to conduct sustained field research on free-ranging dogs and she was invited to publish her observations in Brian Vesey-Fitzgerald's classic English-language text *The Book of the Dog*, a thousand-page tome on every aspect of *Canis familiaris*.[247] She and Rudolph later published their own independent treatise on the topic in 1960: *Pariahunde*, the first comprehensive study of the pariah dog based on their analyses of the behaviors of eighty-one individual specimens. The book explains the concept of the "pariah" dog, describes their origins and the different types around the world, presents advice about the reactions one can expect to encounter when capturing a street dog from the wild, and includes complex discussions about the flight distances of wild versus domestic animals and the differences between domestication and tameness. Their conclusion is that the "pariah" dog "offers a magic key to biological knowledge [and is] a treasure beyond equal."[248]

As it turned out, Emil Hauck's theory that free-ranging dogs were the "missing link" between wolves and dogs was later entirely discredited and Rudolphina's own intrinsic assumptions about what constituted primitiveness and wildness among these dogs underwent similar revision in the later ethological literature. Contemporary thinking about stray, ownerless dogs suggests that all dogs, no matter how variable, descended from the gray wolf in a single domestication event that occurred approximately fifteen thousand years ago.[249]

Rudolphina's related decision in the early 1950s to create a national dog breed for Israel based on these free-ranging dogs was a more fanciful, if no less time-consuming, project. She had already nicknamed these dogs "Canaan dogs," asserting that they, like their ancient human companions the Jews, had been the original inhabitants of the land of Canaan. This was a metaphor she borrowed directly from the radical, anti-religious Canaanite political movement in Israel, an elitist group of cultural Zionists who asserted that the unbroken continuity between modern Jews and ancient Hebrews was based not on Judaism but on shared Hebrew language and culture.

The idea of inventing a new dog breed was not novel.[250] Devoted individuals had been concocting modern dog breeds since the middle of the eighteenth century, a practice that only gained in popularity as dog sports spread across Europe throughout the nineteenth century.[251] For example, Karl Friedrich Louis Dobermann created the Doberman Pinscher in 1897, Max Von Stephanitz created the German Shepherd in 1899, and following in this tradition, Rudolphina created the Canaan dog in 1953. Creating a national dog breed out of the local variety of free-ranging dogs was also not an original idea—in the 1930s, the Society for the Preservation of the Japanese Dog transformed the spitz-like indigenous dogs of Japan into a national breed that exemplified Japanese "purity, loyalty and bravery."[252]

Rudolphina's desire to create a national dog breed for Israel had both a symbolic purpose and a practical function. She not only wanted to create a canine emissary that represented the state of Israel in the world of dog sports, she also wanted to create a framework for conserving and

The Extraordinary Life of Rudolphina Menzel

continuing to study the indigenous dogs of Palestine. The best way to do so, she decided, was to transform them from an evolutionary product of natural selection into a modern dog breed.

Rudolphina certainly knew what was involved in creating a modern dog breed: she had to establish a stud book, write a breed standard, and operate a breeding kennel. Entry number one in the Canaan Dog Stud Book was a particularly handsome male she named Dugma (the masculine Hebrew word for "example") whom she first spotted running around on the outskirts of Kiryat Motzkin. Rudolphina later described how it took weeks to lure Dugma into her yard and even longer before he would allow a collar to be placed around his neck or tolerate being led on a leash. Rudolphina persisted in taming him, however, and proudly recounted how he eventually became a devoted pet and companion who accompanied her on periodic bus trips to Haifa.[253]

Dugma became the enthusiastic founder of Rudolphina's Canaan dog breeding program and sired dozens of puppies with carefully selected mates, each of which Rudolphina registered in the Canaan dog stud book. She made careful notes about each individual's parentage, markings, and place of birth and worked to select the best specimens using her temperament tests. Canaan dogs that were captured in garbage dumps or found in Bedouin encampments, like her best-known brood bitch Dugmona (the feminine Hebrew word for "example"), were registered in the Canaan dog stud book as "wild."[254]

Writing a breed standard for the Canaan dog was not complicated for someone with Rudolphina's vast experience with selective dog breeding. While there were free-ranging dogs all over the region—from the heavy, thick-coated Syrian type in the north to the more "refined," short-coated type in Egypt—the Canaan dog, according to Rudolphina, was somewhere in-between. Initially, she wrote breed standards for two different varieties of Canaan dog, the collie type and the dingo type, which she differentiated primarily by body shape, skull type, and muzzle length. Both types possessed the physical characteristics she believed best represented the ideal qualities of a "sturdy, agile, and adaptable desert dog, totally suited to survival in the difficult climate and terrain of his native land."[255] She considered the dingo type to be "one step down" from the collie type on the evolutionary scale, however, and in later revisions to the breed standard, the dingo type was eliminated.

The final Canaan dog breed standard called for a medium sized dog with a well-proportioned head that has a blunt, wedge shape. The muzzle should be sturdy and of moderate length and breadth. The ears should be erect, relatively short, and broad, slightly rounded at the tip, and set low. The eyes should be dark brown, slightly slanted, and almond shaped. The neck should be muscular and of medium length. The ideal body type is square with well-developed withers and muscular loins. The feet should be strong, round, and catlike. The tail should be set high and look like a thick brush carried curled over the back. The coat should be dense, rough, and of medium length. Males should be approximately fifty to sixty centimeters tall and weigh between eighteen and twenty-five kilograms and have two normal, fully descended testicles. The gait should be quick, light, and energetic. The Canaan dog's ideal character was described as "alert, quick to react, distrustful of strangers, strongly defensive but not naturally aggressive. Vigilant not only against man but other animals as well. Extraordinarily devoted and amenable to training."[256]

To establish and popularize her new breed, Rudolphina founded a kennel she called *B'nei Habitachon* (Children of Security) at her home in Kiryat Motzkin.[257] She then established Canaan dog clubs so that the breed could become better known in local dog shows through demonstrations that proved how useful they could be as guard dogs and how friendly they could be as pets.

Although the Canaan dog failed to become widely popular in Israel (many Israeli Jews too often associated them with dangerous strays) Rudolphina achieved remarkable international success with the Canaan dog breed. By the mid-1960s, the breed was recognized by the International Cynological Federation. In 1965, she began exporting Canaan dogs to a breeder in the United States, and a few years later, Canaan dogs were being bred in England. Gradually, a small but devoted international community of Canaan dog enthusiasts emerged. Some were Diaspora Jews who rallied around the breed by establishing kennels called Mazel Tov Canaan Dogs or "HaTikva Canaans." Some were evangelical Christians who wanted a pet dog that "walked with Jesus" and came from "the Holy Land." Others were "natural breed" enthusiasts who wanted a pet they believed was a more authentic embodiment of the "original" dog.

In 1987 Israel issued a Canaan dog stamp to honor Israel's role as host of the International Dog Show, and in 1997, the breed was admitted to the American Kennel Club (AKC), where it remains approximately the 179th most popular breed and competes regularly in international dog

The Extraordinary Life of Rudolphina Menzel

shows. The AKC's fantastical breed history for the Canaan dog would no doubt make Rudolphina proud:

> The Canaan dog was the guard and herd dog of the ancient Israelites. They were plentiful in the region until the dispersion of the Israelites by the Romans over 2,000 years ago. As the Hebrew population dropped, the majority of the dogs sought refuge in the Negev desert, a natural reservoir of Israeli wildlife. Avoiding extinction, they remained mostly undomesticated, with some retaining a form of domesticity living with the Bedouin earning their keep guarding the herds and camps. Some were also guards for the Druze people on Mt. Carmel. They were finally reintroduced to civilization through the concerted modern efforts of their ancient companions, the Jews.[258]

* * * * *

In addition to founding the guide-dog institute, conducting pioneering cynological research on free-ranging dogs, and establishing the Canaan dog breed, Rudolphina spent the decades after the war writing scientific articles and staying active in the Israeli Dog Lovers Associations as a lecturer, dog show judge, and all-around canine consultant. Even after Rudolph was appointed medical director of the Haifa oil refinery in the mid-1950s, he joined her in many of these activities.[259]

Together, they published a scientific exposé on how domestication affects the relationship between semiwild dogs and cats based on observational research, concluding that domestication intensifies the tensions between the two species and that "only when man interferes consciously in the relations between dog and cat" can the antagonism at the root of this relationship be resolved.[260] They conducted research on the influence of husbandry conditions on the incidence of hookworm infestations.[261] In articles Rudolphina published in European dog sport publications in the 1950s, it was clear she had come to fashion herself as something of a Middle Eastern canine correspondent, regularly reporting on the challenges of breeding pedigree dogs in desert climates and offering advice on how to overcome them.[262] She also became interested in training dogs to detect cancer and sustained a professional correspondence about dog-related matters with those amongst her former cynological colleagues in Austria who had not become enthusiastic Nazis.

Though they never returned to Europe, the Menzels started traveling regularly to America in 1955, where Rudolphina established connections with The Seeing Eye guide-dog program in Morristown, New Jersey and travelled all over the United States to raise funds for her Institute for Orientation and Mobility.[263] In 1962, with the help of American donations, she moved the institute from the old British army barracks in Kiryat Haim to a new campus nearby on a piece of land donated by the Jewish National Fund. She also used her trips to America to popularize the Canaan dog breed by visiting prospective breeders all over the country, and in 1966 she helped established the Canaan Dog Club of America.[264]

An equally important purpose of these trips to the United States was to visit Rudolphina's extended family; her stepsiblings Helene (Waltuch) Boyko and Egon Waltuch had settled in the New York area with their families after fleeing Austria before the war.[265] The cheerful and warm Waltuch clan, full of doctors, scientists, and artists, adored their famous Viennese relatives. Rudolphina was the living embodiment of the Waltuch family's informal motto: don't be a *miesmacher*: a defeatist who gives up easily.[266]

For her professional swan song, Rudolphina was appointed Associate Professor of Animal Psychology and Director of the Institute for Ethology at Tel Aviv University in 1962 when she was seventy-one years old.[267] It had been almost thirty years since she first sat with Professor Tur-Sinai in his office at Hebrew University to figure out how best to translate dog-training terminology from German into Hebrew; now she finally had a faculty appointment of her own. Her academic position not only granted her long-overdue institutional prestige, it enabled her to spend her final years lecturing about the topics she loved: the origins of animal behavior, the principles of animal cognition, and the differences between domestication and training.[268] Despite the fact that her lectures were reportedly highly disorganized, delivered in heavily accented Hebrew, and peppered with German words and expressions that few understood, her courses quickly became so popular that students had to be turned away.[269] She initially assigned readings from Heini Hediger's famous German textbook *The Behavior of Animals in Zoos and Circuses*, but soon after her appointment, she spearheaded an effort to write the first textbook on animal psychology in Hebrew; by 1966 she had found a publisher.[270]

When Rudolph died in 1970 at the age of eighty-two, Rudolphina retired from her professional activities and moved out of their home on Barak Street to the B'nai B'rith assisted living facility on Mt. Carmel in Haifa. She remained lucid and active in her final years and continued

The Extraordinary Life of Rudolphina Menzel

to answer any and all questions about dog behavior, development, and training.[271] She died barely three years after Rudolph at the age of eighty-two, and they are buried side by side in Haifa.

* * * * *

Rudolphina's long, complicated, and eventful life was peppered with triumphs, marred by tragedies, and suffused with ideological tensions. She constantly had to negotiate a shifting kaleidoscope of political, scientific, and cultural considerations in order to realize her extraordinarily ambitious scientific goals and activist objectives. She had to reconcile her bourgeois background with her socialist ideals, her Jewish particularism with her ethics of universalism, her expertise training police dogs with her identification with those they were trained to attack (at least in the European context), and her belief in eugenics with her understanding that genetic traits had to be understood as indicative of individual rather than collective differences. She was often the only woman in a roomful of men, the only Jew in a roomful of Gentiles, or both. She never made excuses for either and usually ended up in charge.

The Waltuch family, ca. 1890s, with Rudolphina at center.
Courtesy of Serena J. Fox.

Rudolphina as a teenager.
Courtesy of the Leo Baeck Institute, NY.

Rudolphina (back, center) with other members of *Blau-Weiss*, the Zionist scouting movement. Courtesy of the Leo Baeck Institute, NY.

Rudolphina with the Theodor Herzl Zionist Club at the University of Vienna. Courtesy of the Leo Baeck Institute, NY.

Rudolphina and Rudolph Menzel during World War I.
Courtesy of the Leo Baeck Institute, NY.

Rudolphina and Mowgli.
Courtesy of the Leo Baeck Institute, NY.

Rudolphina with Rudolph and fellow cynologists in Austria, ca. 1920s.
Courtesy of the Leo Baeck Institute, NY.

With Rudolph's parents in Kleinmunchen.
Courtesy of the Leo Baeck Institute, NY.

Rudolphina (center) at a dog show in Austria. Courtesy of the Leo Baeck Institute, NY.

At the Palestine Institute for Canine Psychology and Training in Kiryat Motzkin. Courtesy of the Leo Baeck Institute, NY.

Rudolphina and Rudolph with Haganah officers.
Courtesy of the Leo Baeck Institute, NY.

Soldier and his dog on a security mission, August 30, 1948.
Courtesy of the I.D.F and Defense Establishment Archives.

Dogs on a landmine mission, November 8, 1948.
Courtesy of the I.D.F and Defense Establishment Archives.

An experienced dog and a dog in training, August 30, 1948.
Courtesy of the I.D.F and Defense Establishment Archives.

Rudolphina (at right) after a search operation in Hadera.
Courtesy of the I.D.F and Defense Establishment Archives.

March of the dog unit on "Defense Day," 1948.
Courtesy of the I.D.F and Defense Establishment Archives.

Rudolphina (left) with Helen Keller (center) and Polly Thomson (right center)
at the opening of the Lighthouse Foundation for Seeing Eye Dogs, 1952.
Courtesy of the Helen Keller Archive, American Foundation for the Blind.

Rudolphina and a student at her Institute for the Orientation and Mobility of the Blind near Haifa. Courtesy of the Central Zionist Archives.

Rudolphina and Rudolph in Kiryat Motzkin, ca. 1950s.
Courtesy of the Leo Baeck Institute, NY.

The Waltuch family descendants, ca. 1960s.
Courtesy of the Leo Baeck Institute, NY.

Rudolphina Menzel, ca. 1960s.
Courtesy of the Leo Baeck Institute, NY.

PART II

Perspectives on Rudolphina Menzel's Legacy

TWO

Rudolphina Menzel's Early Years in Austria

MONIKA BAÁR

IN HER MEMOIR, Rudolphina Menzel explained that one of her earliest childhood memories occurred when she was three or four years old. Someone asked her, "Who are you, little one?" and she replied spontaneously, "I am a liberal." She wrote that her whole family described themselves as liberals, so she assumed that being a liberal "must have been something really good and magnificent."[1] This identification first and foremost with liberalism represents a paradigmatic experience of emancipated, middle-class Jewry in Vienna, which showed a growing degree of integration and participation in politics, society, culture, and arts. Nevertheless, in the late nineteenth century, the parliamentary political power of liberals was broken by emerging mass movements— Christian, antisemitic, socialist, or nationalist—and the emancipated cosmopolitan Jewish intelligentsia came to be perceived as a threat to the hegemony of nationalist middle classes. As Rudolphina's memoir powerfully demonstrates, the general increase in antisemitic sentiments during this period triggered Jewry's gradual alienation from the liberal-assimilationist idea.[2]

Accordingly, Rudolphina had no childhood memories of Judaism, Jewish holidays, or anything comparable. As she recounted: "I have no idea which of our uncles or friends in the family were Jews or not. I do not even remember having heard the word Jew. The Jewish problem for us was nonexistent. We were liberals and nothing else."[3] At the age of four, her mother died and her father remarried. Not even her stepmother brought any elements of Jewishness to the family. She first learned that Jews exist and what that meant at school, where she also learned about the "Jewish problem." Only when pupils in her school were divided into groups for religious education did she learn that she was "mosaic." But this was a liberal school in an affluent district, and during her time there, she did not hear a single antisemitic comment.[4]

Rudolphina mentioned two childhood experiences which were formative for the development of her identity as a Jew. In her religious education class, the teacher explained that the Jews had once lived in a land of milk and honey and led much better lives there than in Europe. So Rudolphina asked the teacher naïvely: "If the holy land is so beautiful and the Jews were so happy there, and if they are not doing well in their current homeland, why do not all we Jews move back there?" The teacher angrily replied, "Sit down you stupid goose." Rudolphina reported that this harsh and unexpectedly angry reply aroused not only her defiance but her curiosity.[5]

Rudolphina noted another ambivalent experience when one day she walked with her elder sister through Leopoldstadt, the district which was allotted to Viennese Jews in 1622 and remained a primarily Jewish area until 1938. Two men approached them in the street dressed in caftans and wearing long beards. They looked at her and seemed perplexed and asked her in a "barely understandable jargon" (Yiddish) if she was the grandchild of Reb Fischl Waltuch. That was the name of her grandfather, who was a great learned man who was known for his altruism. When she answered "yes," she was astonished to see that the two men started to bless her in the street. She felt confused about this scene. On the one hand, she remembered that she felt proud to be a descendent of "kings," but she noted that these *Kaftanjuden* irritated her (*Kaftanjuden* was a pejorative expression commonly used for Hasidic Jews).[6] This episode in the autobiography serves as a vivid illustration of the heterogeneous nature of the Jewish community in Vienna and, in particular, of Rudolphina's vivid perception of the difference between "Eastern" and "Western" Jewry.

Up until around the 1880s, the composition of Jewry in Vienna was mostly "Western": they arrived from Bohemia, Moravia, and Western Hungary; they were typically of middle class in origin and aspiration, liberal in political orientation, and their cultural identity was German. In the late nineteenth century, an increasing number of Hasidic Jews arrived from Galicia, whose language was Yiddish and their loyalty was not towards German-Austrian culture but to the Habsburg Empire. Because Galician Jews in Leopoldstadt were usually poor and did not pay taxes, they also remained excluded from community politics. Their lifestyle

Rudolphina's Early Years in Austria

and living standards embarrassed their "emancipated" counterparts, as the scene Rudolphina described demonstrates.[7]

Rudolphina also soon learned about the Zionist movement. Although children were not supposed to read newspapers, she once found a copy of the paper *Die Welt* at home, which was the world medium of the Zionist movement. She read it in a secret hiding place, and she could not believe her eyes; it seemed that there were many adults who shared the idea that she had in her dreams: to lead the Jewish people back to Palestine. And they did not seem to be deluded. Even more, these people who wanted to recreate the nation were not the pious, godly ones—not the *Kaftanjuden*—but those who for her represented "normal" people.[8] The newspaper also made her realize that although her father was, among other things, a member of the *Deutscher Turnverein* (the German Sports Club), Jews were not German—they were a separate people. To symbolize this realization, she began wearing blue and white ribbons in her buttonhole to mark her as distinct from the German girls who wore black, red, and yellow ribbons. Rudolphina claimed that from that point on, her entire youth was filled with the fight for the acknowledgment that Jews are a nation (*volk*) in equal terms with other nations, and therefore, the Jews are not Germans, Russians, French, or anything else, even when they belong to the German, French, or Russian cultural spheres. From this, it followed that the Jews should strive for a normal national life (*Volksleben*). As she put it playfully: "I have never yet seen an abstract dog in the street, only dachshunds, boxers, or shepherds, so I cannot imagine abstract men, only in their incarnations as German, French, or Jewish."[9]

It was in the second year of the *Bürgerschulklasse* that she was for the first time confronted with a concrete manifestation of antisemitism. One of the stationary shops nearby the school sold caricatures of Jews. She described how she marched into the shop and demanded that the owners remove these images, or else she would convince the pupils from her school not to buy anything from him. At the time, there was a lot of talk everywhere about Karl Lueger (1844–1910), the mayor of Vienna from 1897 until his death and cofounder and leader of the Austrian Christian Social Party who became known not only for his crucial role in transforming the capital into a modern city but also for his demagogy and virulent antisemitism. Infamous for supporting fellow politicians who perpetuated the myth of blood libel and for repeatedly pointing to the alleged Jewish influence in academia and the press, he also influenced Adolf Hitler.

Yet in Rudolphina's family, as in many comparable families, the magnitude of the dangerous potential of expressly antisemitic Christian Social politicians was initially not recognized. It was always possible to find ways to explain away their antisemitism As Rudolphina noted in her memoir: "After all, this was just politics, an attack on liberalism in the first place, and the Jews, as the adherents of liberalism, were the best scapegoats. 'No soup is eaten as hot as it is cooked' (*Die Suppe wird nicht so heiss gegessen, wie sie gekocht wird*), was their common refrain."[10] At the time, the Austrian Social Democrats had conflicts with their Czech members, who separated and created their own movement; this naturally affected the party's unity. In the light of this development, the Jewish leaders' goal to achieve the status of a "nation" within Austria, with equality before the law and a certain degree of self-government, was not exactly met with a positive reception.

At the time, girls were not admitted to the state-run Gymnasia, which served as the entry to university, but some of the private and independent schools admitted women. To prepare for the university entrance examination (*Matura*), Rudolphina enrolled in a Christian Trade School. She noted that her teacher of religious education at this school was surprised when she recited the expected texts not in biblical Hebrew but in the Modern Hebrew spoken in Palestine. In her teenage years, she became more eager to find the company of likeminded young Jews. At first, she tried the *Wiener Jüdischer Turnverein* (Viennese Jewish Gymnastics Club), but she noted that the Jews there did not seem to be engaged with serious ideas. Once she joined the *Jüdische Wanderbund Blau-Weiss* (Jewish Hiking Association Blue-White), she finally found a group of friends and began participating in hiking trips all over Austria.[11]

Blau-Weiss was one of nearly 450 Jewish organizations that existed in the country in the early twentieth century; this number reveals both enormous variety and enormous fragmentation in the Austrian Jewish Community. A great majority of these organizations were secular, and eighty-eight were Zionist.[12] The popularity of Jewish youth organizations was no doubt due to the fact that Jewish youth were increasingly excluded from non-Jewish clubs and groups. In addition to raising communal consciousness and educating the youth, many of these organizations, especially the scout groups, sought to cultivate the love of nature in their members and emphasized the importance of physical work, often in preparation for the members' possible future emigration to Palestine. *Blau-Weiss* catered mostly to more "Western" Jews and was among the most successful: *Österreichische Blau-Weiss für Jüdisches Jugendwandern* was established in 1913, and after merely five years, it boasted twenty-two local groups with 1,200 members. Inspired by the comparable German-Jewish *Wandervogel* groups, it sought to instill into its members a somewhat abstract notion of Jewish humanity rather than a more specific type of Jewish identification.[13]

After the *Matura*, Rudolphina enrolled in the *Technische Hochschule* to study chemistry and natural sciences, indicating on the enrollment form that her mother tongue was German, but she was of Jewish nationality (*Jüdische Nationalität*). She then went on to study for a doctorate at the University of Vienna in chemistry. When her first chemistry professor died, a new professor from Prague arrived who was also a liberal Jew but not a Zionist. Rudolphina used the earliest opportunity to declare her Zionist sympathies, but it did not affect their relationship, and she reports that she became the professor's favorite student.[14]

At the time, Jewish women constituted a disproportionately large percentage of the female population at the University of Vienna, which started to admit women in 1897. Shortly before World War I, they made up forty-three percent of the female students and fifty-nine percent after the war. No data exists for the *Technische Hochschule*, where the percentage of women was comparatively low.[15] Most of these Jewish women claimed German as their mother tongue (although some listed Polish, Yiddish, and occasionally even Hebrew), and the number of Zionists like Rudolphina, who claimed Jewish nationality, was very small.

In many ways, Rudolphina's educational experiences corresponded with those of her fellow Jewish female students, who came from comparable social backgrounds. For families from the enlightened Jewish bourgeoisie, facilitating serious educational opportunities for their daughters was of great importance. Often, Jewish women who obtained higher education were not only able to pursue careers that enabled them to earn a living but they were also able to engage in political life. Yet, the majority of Jewish women studying in Austrian (and German) universities did not possess a strong Jewish identity, and though they received an excellent secular education, they had no formal knowledge of Judaism apart from the religious classes at school. Neither did they adhere to Jewish practices at home. For many of them, Jewishness did not form a crucial aspect of their identity before the rise of Nazism. Another typical phenomenon was that Jewish women in higher education rarely joined Jewish women's organizations, nor did they strive to establish their own associations. There was relatively little involvement in the Austrian women's movement on the part of Jewish women, although, ironically, they were frequent subjects of antisemitic attacks because of their alleged "feminism." Generally speaking, during the time of Rudolphina's student years, Jewish women were more likely to suffer discrimination for their gender rather than their religion.[16]

Rudolphina's decision to join the *Verein Zionistischer Hochschüler Theodor Herzl* (Theodor Herzl Jewish Students Association) at the University of Vienna was unusual, and she became the association's first female member. This was where she met Rudolph Menzel, her future husband. She contrasted this society to the German liberal student organizations which recruited rich Jewish students who became pretentious *Paradendeutsche* ("pseudo"-Germans). When World War I broke out, Jews like Rudolphina and Rudolph felt that it was their obligation to support Austro-Hungary even if they did not agree with its intentions during the war. They hoped that Jews would receive greater autonomy after the war and would receive official recognition as a distinct Jewish nation rather than a minority religious community. After the war, the Menzels settled in Linz where, what Rudolphina called, their "life-long journey with dogs" began.

THREE

Rudolphina Menzel's Invention of Modern Dog Culture in Israel

RACHEL KORIAT

Dr. Menzel was a vigorous, forceful personality and not an easy person to work with, she demanded a lot from us, as she demanded from herself. Our workday was 14-hours long, and when one of the dogs was unwell, we many times spent the night overseeing training and treatment of dogs. Even then I was surprised by the fact that this small, lively woman managed to obtain the trust of the dog trainers. I have no doubt that one of the contributing factors was Dr. Menzel's sense of mission, which she managed to infuse into many of those who came into greater contact with her.
The team morale, I should state, was excellent.

Ephraim Benhar, on working with Rudolphina in the 1940s[1]

Training develops discipline, an undisciplined dog will do as they please on the street and at home, and such dogs give dog handling negative propaganda and add to the number of people who hate dogs. Whereas a well-trained dog provides a successful example to the rest of the dog owners, and will encourage people to train their dogs instead of letting them go wild. Goethe can be quoted here: "a dog with a good education will make even the wise person love them."

Rudolphina Menzel[2]

I FIRST LEARNED about Dr. Rudolphina Menzel from my late father Shmuel Tzvieli-Toretzky, one of the founders of Kibbutz Tirat Zvi in the Bet She'an Valley. He had been one of her students in the canine handlers course she taught in Mikve Israel in 1937. He told me that Rudolphina was an expert in training dogs for guarding and tracking and that she taught dog-training courses to members of the Haganah before the State of Israel was founded. He also said she was a *yekke*, a Hebrew term with slightly derogatory connotations regarding the punctiliousness common to Jews of German origin. That was all I knew of her until I began my archival research and conducted interviews with those who knew her personally.[3]

How did Rudolphina Menzel realize her vision of transposing a modern, Western culture of professional dog breeding and training to pre-state Palestine? To answer this question, I examine a range of primary documents to provide an overview of her major accomplishments during the ten years between her arrival in 1938 and the beginning of the War of Independence in 1948. I draw particular attention to letters and articles in which she outlines the chronic financial challenges she faced and which shed light on the two major public conflicts she became embroiled in during this period.

Shortly after the Menzels arrived in Palestine in 1938, they established a national canine institute at their home in Kiryat Motzkin. Modelled on institutions Rudolphina had worked with in Austria and Germany, the new institute was intended to centralize and supervise every canine activity in the country. It was organized into two divisions: one focused on practical matters of pedigree dog breeding, training, and instruction, the other focused on canine research and scientific studies. These programs were integrated so that studies on canine behavior and health could be applied to have an immediate impact on dog training and care. Assembling these activities under one professional roof allowed Rudolphina to exercise enormous control over all aspects of dog training and research.

To recruit students for her dog-training courses, Rudolphina sent the word out to kibbutzim, moshavim, and other remote settlements to "please contact the institute immediately... and note the specific training you desire. In the event that your dog's traits are not yet clear to you, bring them to the institute," where she would conduct temperament tests on individual dogs.[4] She noted that dogs that were not purebred were eligible for training and "based on the messages we receive, we will arrange the courses to be held in Kiryat Motzkin or in other central locations."[5]

The institute's training courses were organized to teach people to assume various roles as dog handlers, trainers, and instructors. Through them, Rudolphina created a cadre of canine professionals who disseminated her training methods throughout the Yishuv and who were beholden entirely to her: she mandated that anyone who participated in dog-training courses, either at her institute or in other regions of the country, had to pass final completion exams she administered herself; all decisions regarding the granting of completion certificates were hers alone.

The first dog-training courses she offered under the auspices of the institute provided instruction in the three domains that were most crucial for the needs of the Yishuv at the time: mine-detecting, message-carrying, and herding. One of Rudolphina's areas of special expertise was training dogs to detect mines, which drew directly on her understanding of canine olfaction. While still in Europe, Rudolphina had invented a new method in which the mine-detecting dog was trained to detect the scent of disturbed dirt rather than the smell of the mine's metal itself. This method enabled the dog to direct the attention of the handler to the specific location of the mine rather than risk causing itself to explode by actively digging at the mine site, thereby safeguarding the valuable life of a highly-trained dog. It represented a significant scientific achievement and contributed to her international reputation in cynology.[6]

David Efrat from Kibbutz Ein HaHoresh, who worked alongside Rudolphina in the 1940s and later became well-known as the owner of The Spring House breeding kennel, recalled the need for mine-detecting dogs and Rudolphina's methods for training mine-detecting dogs thus:

> The Arabs would boobytrap the dirt roads that led to the fields, and the horse-drawn carts that headed out to the fields would go over these primitive landmines and explode. To detect landmines, Boxer dogs possessed the most suitable character traits, they had low hunting drives, were generally not aggressive and were very fond of eating. The first requirement for training dogs was to ensure that the dog was hungry. We would dig a hole of approximately 30 centimeters, bury a little tin box with a piece of meat in it, cover it, and move away about 20 meters. Then we would attach our dog to an extendable leash that we wore on our belts. We would then loosen the leash so that the dog would learn to run to the dig site, show us where the box was buried, and when we took the meat out of the box, the dog would be "thrilled" to eat it as a reward. We practiced this over and over, until the dog learned to locate the dig site at greater distances and after longer periods of time, always receiving a piece of meat as a reward. Rudolphina claimed that with good terrain and favorable conditions, a dog could discover the dig site up to 4–5 days from the time the landmine was placed.[7]

Efrat also described Rudolphina's methods for training messenger dogs:

> One of the issues Rudolphina tried to solve using dogs was sending messages because the people of the Yishuv lacked communication devices. The problem was how the patrol personnel could let kibbutz members know whether there were ambushes waiting for them in the fields. Rudolphina found a solution using "messenger dogs."
>
> First, dogs used to send messages had to have two owners who would give

orders. Second, these dogs had to have the following character traits: low inclination for hunting (so they won't run after hares or other animals), low aggression, and very social. We worked mainly with sighthounds (salukis) and the local dogs, later known as Canaan dogs. The training method was as follows: we filled a perforated bottle with anise (a strong-smelling substance) so that it dripped. We started with a short distance between the two trainers (around 200 meters). One trainer released the dog in sight of the other trainer and when the dog reached him, the dog received a treat. The dog was then taught to follow the scent of the anise in order to run back to the first trainer, where he received a second treat. In this way, the dog learned to run between the two trainers and pass messages that were attached to a pouch on its neck. We ultimately managed to achieve a distance of three kilometers between the two trainers![8]

Rudolphina made the first attempts to purposefully breed dogs for herding in Palestine. She applied her extensive experience with herding dogs in Europe to the special climate conditions of Palestine. As she began her work, she witnessed the difficulty local shepherds had working with dogs, and therefore started a study intended to find solutions for integrating dogs in the service of herders:

> We continued to have an interest in the development of herding dog training and use. This use has been advancing in our country [Palestine] lately, and we should hope to soon achieve the herding dog culture required for the development of the herding market. In this context, we have gained many valuable options from our psychological investigation. We must introduce an entirely new method in this field, which in other countries has been sustaining itself for a long time already. It seems that many difficulties are posed. The fundamental difficulty is that the Israeli sheep are not yet accustomed to the dog as a herding aid and fears it. The actions discussed herein clearly show us how the purposeful use and the scientific studies are related to our fieldwork. Introducing the dog into the herding market in Israel is the call of the times. But it has also allowed delving into scientific questions and by doing so achieving a new, satisfactory concept in the field of animal psychology and training method.[9]

"התאחדות חובבי ומאמני כלבים בישראל" (החומ"ב)

The Dog Lovers and Trainers Association in Palestine (DLAT)

The centralized canine institutes in Germany and Austria depended on large networks of private canine sports associations run by knowledgeable civilians who supported their work by establishing, regulating, and promoting responsible pedigree dog breeding. One of Rudolphina's first tasks

Rudolphina's Invention of Dog Culture in Israel

was to create similar partnerships between her institute and the three Dog Lovers Associations that existed in Haifa, Jerusalem, and Tel Aviv.

She quickly incorporated these local associations under the umbrella of a new, national organization for canine sports she called The Dog Lovers and Trainers Association in Palestine (DLAT). With the help of Martin Goldschmidt, the chairman of the Jerusalem Dog Lovers Association and later chairman of DLAT, Rudolphina wrote a founding document in Hebrew in which she clearly outlined the goals and activities of the organization:

Objective: To develop and supervise pedigree dog breeding and training in Mandatory Palestine.

The stud book: The DLAT is the sole certified authority allowed to keep and manage the official stud book in Mandatory Palestine, according to its regulations.

The action book: Only DLAT is allowed to supervise all canine activity in Mandatory Palestine according to its regulations and issue official certificates.

Judges: The national leadership of DLAT supervises and oversees the practice, examinations, and appointment of the active judges, the character judges of the dogs, and the judges for the appearance of the dog according to special regulation and a set standard.

Dog shows: The DLAT supervises the dog shows and exhibits according to special regulations and promotes their performance.

Exams: The DLAT supervises and assists in the development of dog evaluations and the various competitions; it determines the rules of the exams and awards practice scores and degree certifications.

Breed standards: The DLAT approves all the accepted international standards concerning dog breeds and is authorized according to international regulations to develop and advance the creation of new continuous breeds that were approved by a certified scientific institution.

A scientific research section for dog breeds: The DLAT will make efforts to advance and encourage the establishment of the department for scientific research of the dog breeds. Those who breed and sell dogs without prior authorization are under no circumstances allowed to be accepted as members by the member associations of DLAT.[10]

Despite the fact that Goldschmidt was the nominal chairman, Rudolphina managed every aspect of the DLAT with a heavy hand and appointed herself Secretary of Honor of the Stud Book and Coordinator of Honor for Breeding and Service Dogs, the two most important roles in the organization. Rudolphina believed that she alone had the knowledge and expertise necessary to create a dependable and efficient system for recording the pedigree of every purebred dog in Palestine and to manage the breed-specific stud books. These detailed records served to certify a dog's origin and were the foundation for legitimizing pedigree dog breeding in Palestine.

The Dog Lovers Associations held informal monthly meetings in neighborhood cafés to encourage dog training, share information, and provide an opportunity for social interaction amongst likeminded people. They also organized local dog shows in which dogs without pedigrees were allowed to take part. Once incorporated under the larger umbrella of the DLAT, Rudolphina worked to maintain the social function of the smaller branches of the Dog Lovers Associations; she recognized the value of creating a sense of social solidarity in which dog breeding was understood as an integral part of a larger and more important movement. But if these associations were going to support the efforts of her new national canine institute, Rudolphina knew that they must do more than just provide an opportunity for likeminded people to socialize.[11]

Under the auspices of the DLAT, Rudolphina created and supervised additional programs. One regulated the buying and selling of dogs, which allowed her to ensure that high-quality dogs could be acquired by those who needed them most for guarding, herding, or mine-detecting, regardless of cost.[12] In publicizing this program, she was careful to note that the goal was not to create a profitmaking venture for selling dogs and emphasized that she stood to gain no monetary benefit from brokering the buying and selling of pedigree dogs.[13]

The DLAT also provided oversight for regular canine classification events, in which newly registered pedigree dogs were formally admitted to their local associations. Finally, the DLAT provided the all-important framework for canine temperament tests, which of course were entirely dependent on Rudolphina's extensive expertise in the field. As she noted, "without the existence of the training and character exams, it would not have been possible to build a proper professional set of service dogs."[14]

Her efforts were rewarded in September 1945, when the DLAT met the stringent criteria for membership in the International Cynological

Rudolphina's Invention of Dog Culture in Israel

Federation and received international recognition from the Kennel Club of London, which registered it as an affiliated association.[15]

Research on Canine Diseases and Treatments

In the German and Austrian canine institutes Rudolphina knew so well, research on canine diseases and treatments took place alongside research on canine behavior and development. Rudolphina attempted to establish the same sort of framework under the auspices of her new institute and wanted this kind of work to be conducted under her supervision. She saw herself first and foremost as a scientist who dealt with practical research and its applications to all the questions that pertained to the dog's life. To her, there were no divides between behavioral-psychological disciplines and physiological-sanitary disciplines.[16] So she reached out to local researchers and veterinarians who were actively engaged in the study and treatment of rabies, leishmaniasis (a severe skin disease), and salmonella: diseases which were rampant in the country and affected many dogs.[17] For example, in a 1939 letter to Dr. Beham at the Institute Pasteur in Tel Aviv, Rudolphina wrote:

> My very esteemed colleague sir, we are determined to carry out a prophylactic vaccine for all the service dogs used for guarding, defense, and more. Additionally, there is a need to vaccinate a few dogs who during guarding duty received some sort of wound. Regarding the matter of injured dogs, please let us know the cheapest price possible. We hope he will recognize our matter as being an important issue for the safety of the Yishuv and not as a private matter.[18]

To combat leishmaniasis, Rudolphina worked with closely with Dr. Adler at the Parasitology Institute at Hebrew University:

> On the subject of these important questions we are in constant contact with the relevant Hebrew University departments, and with the Veterinarian Doctor Association in Mandatory Palestine regarding the pandemics that endanger dog care. Here we emphasize the various types of works, canine typhoid (Stuttgart Disease), rabies, leishmaniasis, trichophyton, and animal skin treatments. This list shows us the connection these problems have with the questions of human hygiene. This connection is very important considering the shared life of man and dog. As to the prevention of rabies, we wish to provide a general vaccine for all employed dogs in the country, of course in partnership with and with the assistance of the veterinarian doctors. For that purpose, we are still missing some statistics,

because it seems to us that the local attempts made regarding this issue do not align at all with the attempts made abroad... regarding the war on ticks we are trying a series of different methods. For this purpose, we are in contact with the national pharmaceutical industry to perhaps manufacture a novel, cheap means suitable for the biological needs of the dog. We paid consideration to the question of nutrition and its doctrine, after many lengthy attempts we now have the ability to make dog food for our own use, by self-manufacture. This product is cheaper, more restoring, and more nutritious than that sold in stores of foreign manufacturing. We think to establish such an industry to sell domestically, not as a sport. We have already negotiated this matter.[19]

To address the problem of salmonella poisoning amongst dogs, Rudolphina established a productive partnership with Dr. Reitler from the government hospital in Haifa, which led to great success in the battle against this disease:

We can now share that we have reached a solution regarding one disease that has claimed many victims throughout the country. With the vigorous help of Dr. Reitler from the government hospital in Haifa, a new cause of Salmonella was discovered. We also succeeded, at least partially, in finding out the contagion pathway of this disease and its relation to other pests. We have already started with initial attempts to vaccinate dogs at the Institute. We hope that the benefit of the vaccination will be satisfactory. Based on our experiences in this field, we wish to initiate the vaccination of all service dogs in Israel[20]

These partnerships with local researchers were not mirrored in her relationships with local veterinarians, however (an association for veterinarians in Palestine had been founded in 1922).[21] Only six months after the Menzels settled in Kiryat Motzkin, Dr. Rodolfson, a veterinarian in Hadera, wrote the following to Rudolphina in early 1939:

In many cases you did not respect the justified authority of animal doctors. I personally know of two such cases—your efforts on the matter of a rabies-prevention vaccine and the matter of pharmaceutical manufacturing. As much as your efforts on both issues are admirable, you did not bother to involve any sort of authorial force in the matter. In the current situation, the veterinarians must look at the actions of your institute as reaching beyond the limits of your jurisdictions. I can do nothing more other than offer you collaboration with the committee. Please send me all material pertaining to veterinarian issues and refrain from doing anything related to veterinary issues until further notice from us. I am convinced this is the only way to serve our interests.

Rudolphina's Invention of Dog Culture in Israel

I do not comprehend how you took on this difficult and heavy work without collaborating with the local animal doctors, *aside from the issue of dividing work according to fields one cannot dismiss the agreement and goodwill of animal doctors* for an encompassing, successful execution of the vaccine issue [emphasis in the original]. Your behavior is not only tactless but also harms the work. Only collaboration with the veterinarians can achieve the result you desire.

The second case where you went above the head of the veterinarians and harmed the interests of the issue is your negotiations with a pharmaceutical company. Even had the company not turned you down due to the lack of collaboration with veterinarians, manufacturing the drugs for the dogs would have been an inevitable failure. The veterinarians would have forced the company to halt production, among other reasons because drugs at the hands of the uneducated could harm animals.[22]

Rudolph wrote the reply to Rodolfson, even though the attack had been directed at Rudolphina, and tried to resolve the conflict amicably:

You are completely right. From our own perspective, we might have been in a too-romantic mood after fleeing from the grip of a terrible enemy and hoping to realize our life's dream. We were too influenced by the feeling of immigrating to a homeland. Additionally, we came with a mission of erecting from nothing an organization that has yet to exist, meaning, without the need to shoulder our way within the dire economic situation of our country. Before everything, a general comment—we do not intend in any way to injure the interests of the veterinarians who operate here, and beyond that, we are connected to the veterinarians here by multiple ideological goals and dear connections of friendship, so such intentions are distant from us.

Since 1922 we have dealt with the matter of a rabies-prevention vaccine. We published on the subject and corresponded with acclaimed professionals around the world, among them Claude Frand [a well-known Swiss researcher who studied rabies] since 1934. The processing and scientific consideration of this important matter is part of the institute's agenda. On this matter, we have not done anything behind the backs of the veterinarians, or something that went against their interests. So it is not really clear to us what we stand accused of. Of course, the practical execution of a prophylactic vaccine is under the purview of a veterinarian, if there is one in the area. No vaccine has been given to date by someone other than a veterinarian or a physician. I am racking my brains and cannot find any reason for the sudden anger that has appeared regarding this matter. There must have occurred some sort of misunderstanding that remained unresolved.[23]

Rudolph clearly felt compelled to remind Rodolfson of the Menzels' professional authority in matters of canine disease and his tone indicates that they were clearly offended by Rodolfson's implication that they were acting from negative, selfish, or financially-driven motivations. These tensions, once exposed, were never fully resolved, and research on canine diseases and treatments in pre-state Palestine remained decentralized. Rudolphina did not succeed in her efforts to establish herself as the supervising authority over all research programs on canine diseases by incorporating them into the research arm of her institute. As a result of this initial fallout, the Menzels worked assiduously to improve their relations with local veterinarians in subsequent years.

The Institute's First Year of Operations

Rudolphina prepared a meticulous report on the institute's specific activities and range of work that characterized the first year of operations. In 1939 she noted that she conducted 120 supervisory activities outside the institute and 407 within the institute; visited 63 settlements, performed 235 canine evaluations, spent 109 days teaching courses, gave 54 lectures, and showed 6 dogs for education and promotion.[24]

In-person consultations were an extremely important part of building the institute's infrastructure because through them, Rudolphina nurtured and maintained communication with her students who were now training dogs around the country. The activity was carried out in two ways: visits by Rudolphina to different locations throughout the country and visits by dog owners to the institute.

Rudolphina described how the institute's growth as the regional center for canine advice had already grown in their first year of operation:

> This role is increasingly expanding as the public's understanding expands with regard to the importance of the dog and its service. 1,433 letters were sent by us throughout the year. People have come to see us with dogs for different purposes at 101 locations. Outside the abovementioned locations, 149 locations have addressed us without dogs for counselling in person or in writing.

She also described her optimistic future plans and objectives:

> The institute will serve as a center for professional canine advice by:
>
> 1) Organizing visits to all settlements using guard dogs for security.
> 2) Examining all service dogs and correct mistakes in handling.

Rudolphina's Invention of Dog Culture in Israel 101

3) Visit settlements requiring dogs but not yet using them.
4) Find out the service possibilities and the requirement for different types of dogs by trials on-site instead of using the institute's dogs [Rudolphina would lend the settlement dogs from the institute in order to use them to check the suitability of the dogs for the settlement's special needs].
5) Answer questions in person and in writing, and clarify unclear situations by visiting and examining them on-site.[25]

Rudolphina succeeded in achieving many of these goals in the early 1940s, but she didn't realize a key part of objective five until April 1944, when she published the first issue of *Hakalban* (The Dog Handler), a monthly magazine intended for members of the DLAT. It was published in two languages, English and Hebrew, and the chief editor was Rudolphina herself. The magazine had several hundred subscribers.

The use of *Hakalban* as a platform to provide professional advice regarding all aspects of dog care, breeding, and training suggests that Rudolphina was not satisfied with only providing advice through personal correspondence and saw the need to found a publication to disseminate professional knowledge written by herself and other specialists. Rudolphina's purposeful integration of veterinarians into the magazine simultaneously legitimated the advice they provided and indicated their active cooperation with her institute; it also symbolized her attempt to mend fences after her initial disagreements with them.

In addition to writing a recurring feature section entitled "This Month in Dog Handling," Rudolphina wrote short articles primarily concerning dog education, behavior, and training. In the June 1944 edition, for example, Rudolphina's articles were titled: "What is the purpose of rearing pure-bred dogs?" and "At what age should we start working with a young dog?" In the June 1945 edition, her articles included: "What to do if a dog wants to bite during feeding, even its own owner?" and "How should dogs be handled in a fire?" Occasionally she wrote articles on canine health as well, such as: "How to treat sand flies that transmit skin diseases."[26]

The articles by local veterinarians offered practical advice about canine physiology and disease. For example, a Tel Aviv veterinarian named Dr. Yeruslavski wrote an article entitled: "Do not use 'Kettle-deep'" (a dangerous disinfectant that could result in death). In the same issue, Dr. Caspi, another Tel Aviv veterinarian wrote: "Dog parasites that might transfer to humans."[27]

Each edition of *Hakalban* included a final section called "Questions and Answers." This last section provided Rudolphina with an opportunity to give written advice to specific questions from readers and was one of the most important tools by which Rudolphina established her status in the field. There were often numerous questions about dog training, behavior, and care, such as:

- What are the preferable education methods for different types of dogs?
- When and how are dog temperament tests performed?
- Could "any ordinary person" train a tracking dog by themselves?
- How should different kinds of behavior problems be handled (barking dogs, dogs who refuse to eat, etc.)?
- How to handle dogs "affected by the madness of love"
- Do dogs dream, and how can this be detected?
- How to handle a dog giving birth, etc.[28]

Sometimes the questions were more halakhic, such as: Is a Jew is allowed to train a Boxer dog owned by an Arab? Other questions concerned dog diseases in general and rabies in particular.

Through the pages of *Hakalban*, Rudolphina managed to form more personal connections with the dog owners in Palestine, who saw in her a professional authority of the highest rank and sought her guidance on all questions regarding dog care, handling, and training. She used *Hakalban* not only to disseminate practical information but also as a platform to present her academic knowledge in the fields of dog psychology and behavior, thus establishing her authority in all canine-related matters.

In order to run her institute, organize the DLAT, oversee the stud books, conduct research, and publish *Hakalban*, Rudolphina needed money, and she had to fight constantly to secure financial support for all of her activities. She wrote numerous letters to the Yishuv leaders, primarily Eliezer Kaplan—who was the treasurer of the Jewish Agency in the late 1930s and early 1940s—in which she described the economic and practical difficulties she faced concerning the everyday operations of the institute. In a September 1942 letter she wrote:

> The purpose of the memo is to finally discover someone high up who will listen, and the help we need, without which we cannot sustain our institute The institute was established four years ago based on a small but sufficient budget for fulfilling the roles on a small scale. When we emigrated we were promised, as you know, a suitable institute and a satisfactory budget so we would be able to develop and sustain the dog training field in our country. The outbreak of the war [World War II] and the death of our friend A. Ruppin prevented the immediate execution of this operation. Furthermore,

Rudolphina's Invention of Dog Culture in Israel

we were assisted by the very large reserves taken out of our own previous assets (such as typewriters, office resources, papers, furniture, bedding, and undergarments for the institute's employees, canned food for the dogs, training equipment, and more). All these were our own assets, which have now broken down in parts or have been used up.[29]

She repeatedly states her disappointment that promises made to her and her husband were not kept, both regarding the employment of her husband as a physician for a health maintenance organization or Hadassah hospital and regarding budgets promised by various entities.[30] She goes on to explain how these difficulties made it almost impossible to continue running the institute and that while the institute was asked to do more and more dog training, particularly to supply dogs to the British Army during World War II, its budgets were repeatedly cut:

> When, around 18 months ago, the aforementioned budget was cut by half, we were promised assistance on both sides. First, by providing my husband work at a health maintenance organization or at "Hadassah" second with support from agricultural institutes. All those promises proved false. My husband's work at the HMO is nearly nothing, and on the other hand, we further burdened the institute's tasks with additional work, that is, working with herding dogs. We did not receive a single lira yet on this side, and moreover, we had special travel expenses for this job.[31]

Rudolphina goes on to describe the direct and indirect economic benefit the Yishuv would see from professional, efficient employment of dogs for settlement, guarding, and herding, as well as all the benefits the institute generated in the past by for providing guard and tracking dogs to the Yishuv.

> One cannot ignore the financial profitability of such efforts. The prevention of theft of produce, sheep and goats, cattle, poultry, and so on, the assistance dogs provide in taking over land, are facts that cannot be ignored based on the recent attempts in the annals of Jewish pioneering... Sparing one murder trial brings in for the Yishuv sums that could be used to assist the matter of the dogs for many months. And no one can rightly estimate the number of murder trials prevented by the presence of a dog. Indeed the great tragedy of the dog's role is that you usually only feel the benefit when it is not there.

In other words, guard dogs served to prevent the loss of life of Arab thieves by enabling Jewish sentinels to catch or chase them rather than kill them, thereby forestalling both expensive lawsuits as well as the demand for blood vengeance. Rudolphina ended her letter to Kaplan by itemizing the funds needed to run the everyday operations of the institute and with a request to resolve the legal-public standing of the institute:

> In order to correct this unbearable situation, I ask for the following amendments:
> A) A one-off payment for the execution of the most urgent jobs needed (winter-proofing, fixing or replacing training equipment that ran out or broke, the establishment of a quarantine area, and more) at the minimal sum of 70 Palestinian pounds.
> B) Restoring the institute's budget to at least the previous status (50 Palestinian pounds a month).
> C) Efficient assistance from the large institutes in our effort to obtain support from abroad for the independent establishment of the institute.
> D) Regulating the institute under the patronage of one of the public legal institutes of the Yishuv.
>
> We therefore ask for support towards the establishment of the institute at a sum of 3,000 Palestinian pounds, *in the form of the usual loan extended to national entities* [emphasis in the original]. We so hope that you will finally approve the monthly allotment of 50 Palestinian pounds so that the institute can be run according to its needs.
>
> We note that an organized, well-equipped institute will serve as a center for scientific studies in animal psychology, and *their acclimation and domestication for the entire Middle East* and all the parts of the British empire. Such an institute could positively influence and help with Zionist propaganda and do a lot to help increase the scientific renown of the Yishuv.[52]

Rudolphina's entreaties fell on deaf ears. On November 11, 1942, Eliezer Kaplan replied:

> I read your letter where you ask for 3,000 Palestinian pounds for building the institute. I am very sorry that I cannot take on your request, considering the current situation of our budget. *We do not know of any such promise made by the settlement division in this field. We value your work*, and we would advise you to approach a banking institute and ask for a five-year loan [emphasis in the original].[53]

Kaplan's negative answer was very disappointing and frustrating to Rudolphina because it meant he was unwilling to recognize the institute as a national entity but rather advised Rudolphina to take out a regular loan from a private banking institute, as though the institute was a private, for-profit business rather than a national body intended to contribute to the settlement and defense of the homeland.

Rudolphina and the British

Rudolphina often wrote about the differences between English culture, which showed a positive attitude and appreciation towards dogs, and "primitive" Jewish culture, which lacked any understanding of the dog and its possible contributions to security and defense. She developed relationships with officials in the British Mandate both informally at dog shows in the early 1940s (at which she and various wives of British Mandate officers would serve as judges) and more professionally through her various dog-training activities. The collaboration between Rudolphina and the British solidified during World War II when Rudolphina became virtually the sole supplier of service dogs to the British army in their North African campaign. She recalled:

> As the war broke out, the English already knew the true value of the dogs, and so they approached our institute, which they were already familiar with and which was then the center for dog training in the country, and asked for efficient dogs to help. After consulting with the Yishuv authorities and particularly Moshe Sharett, every possible aid was offered to the English armies. This decision was not amongst the easiest, as the Englishmen's war against Hitler was our war as well, but on the other hand, we were forced to give up some of our excellent dogs, but there was no other way. The English could not obtain suitable dogs from anywhere else. Europe was under Hitler's grip, and our dogs were suited to the present need not just due to their attributes but because they were acclimated to the region. They received from us 400 dogs, meaning most military dogs in the Middle East were Mandatory Palestine-raised. Alongside the dogs they received from us, we also provided advice on demand, and at this opportunity (after consulting with Moshe Sharett), the English were given the method of training dogs to detect land mines. From Mr. Sharett I received the complete authority to negotiate dog training-related business with the English according to my own decision.[34]

The Urabin Affair

I conclude with an overview of the so-called Urabin affair of 1947, which proved to be a catalyst for the demise of Rudolphina's institute for canine research and training.

Shlomo Urabin was one of the senior dog trainers who worked with Rudolphina for many years as part of the Dog Lovers Association in Tel Aviv. He retired from the organization in July 1947 due to professional and personal disagreements with Rudolphina, and a few months later in December 1947, he published an article in the popular daily newspaper *Davar* in which he attacked not only her but also her beloved Boxer breed:

> "The dog in the service of defense" by Shlomo Urabin
> One cannot imagine a frontier force guarding the borders without trained dogs. Just as much as an innovative police force cannot be imagined without tracking dogs. However, Israel is the only country where security forces use only one breed of dogs, that is the Boxer, even though this breed was definitely disqualified for use by the defense system. And only in Israel does the training period last 5 days, whereas in other countries it lasts a whole year. Therefore, it is no surprise that we have the highest mortality of dogs during their service.
>
> The Jews in Europe were not active in raising and training police dogs. Ten years ago, the first dog breeders came to Israel, and they just happened to be Boxer breeders, and have introduced this breed here, even though it has failed in the service of police around the world. The Boxers do not have perseverance. They are also not suitable for the climate in Israel. Whereas other breeds, such as the German Shepherd as well as the Airdale Terrier withstood the test in other countries and are worthy of being raised in Israel as well. In Austria, all dog breeds were reviewed in terms of efficacy for the police, and it was determined that the German Shepherd is the best of all. The Swiss Frontier Force also used the German Shepherd almost exclusively. The experts in Switzerland determined that Boxers cannot accompany the policeman in their action in difficult climate conditions. An expert from the police administration in Munich has reported that the police in Bavaria has not used Boxers for 15 years. In England, both the army and the police are using only shepherd dogs, as is the rule in America and Italy as well as Egypt, where the climate is similar to that in Israel. The time has come to quit raising Boxers and instead we should introduce a breed appropriate for the job of guarding, in addition to extending the training period. Neglecting this field could cost a high price.[35]

Even though the name Rudolphina Menzel was not mentioned in the article, it was clear that Urabin was slandering Rudolphina and accusing her of being unprofessional and motivated by narrow interests on subjects related to matters of breeding and training service dogs. At the same time, Urabin approached the leadership of the Jewish Agency and presented himself as the professional authority for training service dogs.

Rudolphina not only felt betrayed by one of her senior apprentices, she had to defend against his attack on her prestige and professional integrity; a clash reminiscent of the conflicts between Rudolphina and the veterinarians was unleashed. Her first response was to activate her canine networks. By 1947, Rudolphina had trained an extensive group of dog handlers, regional canine coordinators, and dog trainers who considered her the ultimate professional authority for any matter and issue pertaining to breeding and training service dogs. Rudolphina personally solicited letters to *Davar* from her acolytes, evidenced by this letter to Leo Frederick in Tel Aviv on December 12, 1947:

> Have you read in *Davar* last Friday that article filled with silliness and lies? The union has decided to organize a stream of protests and I want to ask that you as well write a letter. It is important to emphasize: That it is an outright lie that the dogs only study 5 days. On the contrary, the 5-day courses are intended to provide the handler with the theoretical and practical basics. On that basis he must later work with the dog with the help of local experts. Only after many months can he undergo an exam. *I ask that you immediately write that you will demand they contact the service dogs handler union and not publish an article by a dissident* [emphasis in original]. Maybe you know other people who can also write?[36]

She also aimed her wrath at *Davar*, which instead of representing the views of the central and legitimate canine organizations, decided to give Urabin a public platform to air his challenge to Rudolphina's authority. Finally, on August 5, 1947, she wrote directly to the Jewish Agency to defend her name and reputation:

> We think The Agency is aware that a man called Urabin is trying to disturb the peace of the dog training organizations in the country, for personal reasons. As we read in a well-known newspaper, this man bothered The Agency again with "memos" and "professional opinions." Today we are tasked with the responsibility for the development of service dogs as part of the security of the Yishuv.[37]

She went on to demand total authority over all canine operations in Palestine:

> We wish to emphasize that in the interest of unified, beneficial work in the field of dog training in our country, it is ill-advised that the clerics of the civil department of the agency answer questions on the matter of breeding and training dogs, without first passing the matter on to our organization, or to the institute in Kiryat Motzkin. We kindly ask that all matters of dog training will be passed on to this organization (the Service Dogs Handlers Union in Palestine), or to the aforementioned institute, such as when a person looking for first aid assistance will be directed to "Magen David Adom" or the matter of extinguishing fires will be directed to the "fire

services." Further proof is that during the war, Mr. Sharett handed over any dog-training related decision, as it related to the relationship with the allied armies and the local police, to the administration of Dr. Menzel's institute, and only matters of discretion prevented the addition of the institute to one of our institutions.[38]

Despite her efforts to discredit Urabin, the whole affair contributed to the erosion of her complete control over all dog-related activities in the country. After the State of Israel was established, her institute was dissolved and her role in the canine corps of the new Israeli army was downgraded to that of a part-time advisor. Nevertheless, during the intense ten-year period from 1938 to 1948, Rudolphina recorded an impressive string of achievements and effectively transposed an entirely new cultural repertoire of dog breeding, training, and care to Palestine. She almost singlehandedly established civilian dog sports and created a broad range of associated professional activities. She not only developed these pioneering ideas, she knew how to put them into practice.

FOUR

Canine Zionism: Rudolphina Menzel and Working Dogs in Mandate Palestine

BINYAMIN BLUM

FOCUSING ON Rudolphina Menzel's role in introducing dogs into the Haganah, this chapter explores the role that dogs played in the broader Zionist agenda before the establishment of the State of Israel in 1948. Both practically and symbolically, dogs were significant to the Zionist movement. From a practical standpoint, dogs fulfilled important tasks in Jewish settlements. They served as guards, trackers, and messengers in the new settlements established in Palestine. Furthermore, knowledge of tracker dogs' modus operandi helped underground Zionist operations evade detection. Symbolically, the Zionist movement consciously broke with the diasporic Jew's stereotypical fear of dogs. Reforging the relationship between Jews and canines signified a return to nature and to the land. Zionism's relationship with dogs also marked Palestine's Jewish settlers as distinct from their Arab neighbors, who were frequently portrayed—often inaccurately—as fearful and hateful towards dogs. Through Menzel's efforts, dogs were also used to strengthen the bonds between Zionism and the British government: by providing highly trained dogs during World War II, primarily for mine detection, the Zionist movement proved its usefulness in the war effort.

Dogs and the "New Jew"

By the beginning of the twentieth century, Jewish attitudes towards dogs were undergoing a conscious rebranding, at least in some circles. Through a stronger bond with land and nature—including animals—the Zionist movement sought to forge a new Jewish identity. Dogs played a key role in this plan. The relationship between dogs and the Zionist agenda was made explicit in Rudolphina Menzel's writings. In the foreword to her 1939 manual *Dog Education and Training*, she drew on stereotypes concerning the diasporic Jew's irrational fear of dogs to explain the key role of dog-training in Zionist ambitions to forge the "New Jew":

> We [the Jews] are predominantly urban dwellers and the descendants of urban dwellers. The blood that runs in our veins is that of generations that spent their lives in urban suffocation and of generations before them who never left the ghetto alleyways. They were far from nature, far from the land, far from animals.
>
> All animals were alien to them, the dog being the greatest stranger. The dog belonged to the world of the gentiles. He served as the oppressors' companion and accomplice; at the landlord's command he would attack and chase away the Jewish peddler; the dog was the companion of rulers who determined the Jews' fate, good or ill.
>
> Yet before the ghetto dwellers there were other generations of free peasants, men of agriculture and herding, of war and of hunting. In those days our ancestors lived in harmony with nature and their lives reflected nature. In those times the dog was a companion who assisted our people.
>
> Our national revival movement is erecting bridges to antiquity, skipping over many generations and linking us more tightly with natural life forgotten for centuries. In our treatment of dogs we must also pass over many generations and strengthen our ties to ancient traditions of a nation of shepherds and farmers in the ancient land of Israel.
>
> The dog is like no other species. All species cling to their own. The dog alone has transcended his species to join man and become his companion and helper. Let him be a companion and assistant in the rebuilding of our land.[1]

Assisting Jewish Settlement

Dogs played a key role in Jewish settlement before statehood, especially after 1936, when the British began strictly regulating weapon possession. Dogs then became crucial in protecting the new settlements. Of the many psychological benefits dogs provided to Zionist settlers, guard dogs instilled particular confidence, as illustrated by the description of the 1945 establishment of Gezer, a kibbutz between Jerusalem and Tel Aviv. One of the settlers, writing under the pseudonym "Gideon," described the role that Rudolphina's trained dogs played in facilitating the establishment of the kibbutz. The fifteen founding kibbutz members included two Boxers. The one-year-old Sabra had just completed training at Rudolphina's institute in Kiryat Motzkin. The other, Lapid (torch), was on loan from the institute specifically for the purpose of guarding the settlement. Gideon described how on the day of establishing Gezer, visitors and guests from neighboring Arab villages came to congratulate the settlers, with the dogs inspiring "respect." But the dogs also inspired a sense of superiority and security: not having yet received weapon permits, the only weapons in the hands of the settlers were these two trained Boxers, who would join the guards when executing their nighttime duties. Gideon described how the Boxers could detect visitors (and intruders) from a distance of up to two kilometers away. Settlement, he said, would have been impossible had it not been for their services. Gideon claimed that the Arabs believed that these two Boxers were specially trained hyenas held by the settlers.[2] In yet another unflattering stereotypical depiction of the kibbutz's Arab visitors, Rudolphina even went so far as to claim that her dogs could reliably distinguish Jews from Arabs based solely on their body odor.[3]

Rudolphina insisted that her dogs were trained only for defensive purposes, not to attack. In one instance, a Jewish settler was attacked by two Arabs. The Jewish settler summoned his dog, who had been trained by Rudolphina. The dog severely injured one of the attackers and the British authorities prosecuted the dog owner for unlawful possession of an assault dog. But at trial the dog was brought to court, where he calmly walked around the courtroom and affectionately licked the hands of the judges. The defendant was acquitted. One of the judges reportedly told the defendant that he would be happy to adopt any of that dog's offspring.[4]

Dogs and Anglo-Zionist Relations

In addition to re-establishing Jewish bonds with nature, Rudolphina regarded dog training as a method for strengthening the often-strained relationship between the Zionist movement and His Majesty's Government by assisting the British war effort. In a 1942 memorandum to the Jewish Agency, Rudolphina observed that "The desire for dogs as contributors to the war-effort in different tasks has increased... during the past year." To this end, she noted,

The use of dogs has proved necessary in the following activities of the Middle East Forces:

1. In the Military Police (especially in Egypt).
   Fighting dogs against spies, watch-dogs for guarding ammunition dumps and preventing sabotage.
2. In the Buffs.
   For guarding military and prisoners' camps, aerodromes, stores as well as for detecting and fighting paratroopers.
3. In the Coastal Guard.
   Watch dogs and fighting dogs against suspected people and paratroops, perhaps also message carrying dogs.
4. In the Palestine Voluntary Force.
   Mainly watch dogs.

Rudolphina therefore proposed in Palestine "a mobilization of the dog-material, similar to that which is being arranged in England now." With her staff of experts, she offered to "test the dogs and direct them to their proper use."[5]

Beyond the practical assistance the dogs might provide to the war effort, Rudolphina noted other ways in which the dogs might strengthen Anglo-Zionist relations. She observed that sport often brought nations together, noting: "the good relations between the English and the Arabs are strengthened by English interest in Arab horses and by Arab expertise in horse breeding. We may tilt this unfavorable balance by successful and model use of good dogs, by breeding such dogs and by establishing successful dog sporting in Palestine."[6] In a secret addendum to her memorandum, Rudolphina noted that many Jewish swindlers had taken advantage of the high demand for dogs: despite the risk of bringing disrepute upon the Zionist movement, they provided inferior dogs to the war effort.[7] It was therefore imperative that proper efforts be made to supply the Allies with well-trained dogs.

Soon thereafter, Rudolphina did precisely that. In 1942, as German forces advanced towards Egypt, British authorities approached Rudolphina with a request for assistance in combating what Rudolphina would later characterize as "Arab sabotage."[8] The British asked particularly for trained Boxers, Alsatians, and Dobermans. With the blessing of high-ranking Zionist leaders including Moshe Sharett, Rudolphina began holding "dog drafts." She provided the Allied forces with nearly 400

Working Dogs in Mandate Palestine

dogs, mainly Boxers and Alsatians.[9] By her account, most military dogs employed in the Middle East originated from Palestine.

The dogs Rudolphina trained in Palestine had a marked advantage over the dogs already in the Allied forces' possession and use: they were far more familiar with the local terrain and conditions. The extreme climate of the region presented particular challenges to the working dogs. For tracking dogs, the sandy soils and intense heat meant that "a track is often no more workable after one hour." The sun and dryness proved particularly harmful to the durability of tracks. There were other complications as well: "The poor vegetation, the many thorns, and the many grazing animals make the work of the dog extraordinarily difficult." When working during the hottest hours of the day," Rudolphina noted, "the chances for success are comparatively small."[10] As the Germans advanced in North Africa, European- and North American–trained dogs proved less adept in local conditions.

Rudolphina's dogs provided various military services: tracking, guarding, detecting and rescuing casualties, and carrying communications and ammunition. They were also trained to detect and chase "native thieves and bring them to ground until they can be arrested by the Military Police."[11] In this, the Boxer proved particularly effective, as their "Bulldog breeding gives [them] the toughness and tenacity." One Boxer named Simi reportedly chalked up as many as sixty-eight arrests to his credit. The Alsatians proved more effective in guarding, as they could "see, hear, or smell a native thief when he is hundreds of yards away."[12]

But most notably, Rudolphina had trained her dogs—primarily Boxers—to detect land mines.[13] As the German military transitioned towards using plastic-cased, antipersonnel land mines, metal detectors were no longer effective in mine detection. Rudolphina had therefore trained the dogs to seek recently tilled soil to detect mine placement. Dogs became the most reliable mine-detection technology at the time not only in the Middle East but in Europe as well. However, the Middle Eastern conditions still proved challenging: "with strong wind, especially on sandy soil the dog can, in the direction of the wind, detect a mine up to several yards from [its] place."[14]

In supplying the Allies with trained canines, Rudolphina required that British officials promise her that the dogs she trained would never be used against Zionist settlers in Palestine. This was a promise that, according to Rudolphina, the British kept until the very end of the Mandate. When full-blown hostilities between the British and Zionist movements resumed in 1945, the British did not use Rudolphina's dogs to track down members of the Haganah and the Irgun or their weapon and ammunition caches (known as *slicks*). Instead, they used European and South African trained dogs for these purposes.[15]

Assisting the Palestine Police

Rudolphina also assisted the Palestine police in crime investigation. Especially in Palestine's northern region, the Palestine police had difficulty quickly deploying their Doberman Pinschers from their kennels in Jerusalem (in time, this would lead the Palestine police to establish a second kennel in Afulah). Rudolphina would often be the first on the scene with her dogs Maggie and Blitzie. Recalling a 1939 murder case, she described how she arrived a full three hours before the Palestine police but lamented that she was not allowed to start the search until the Criminal Investigation Department's (CID) Dog Section had arrived at the scene for fear of corrupting the tracks. This was particularly regrettable because it was a cold, damp day, and the tracks would have more easily been followed when fresh. Following the CID, Maggie and Blitzie actually tracked down a different trail, highlighting the significance of dog handlers and the techniques involved in training tracker dogs.[16]

In some cases, however, the Palestine police allowed Rudolphina to proceed on her own. In a January 1939 homicide investigation outside of Kiryat Bialik (a suburb of Haifa), Rudolphina was alerted at around 3:30 p.m. to try and follow the tracks at the crime scene. Presumably, the Palestine police would not be able to bring the dogs before dark, which would seriously impede their usefulness. Rudolphina rushed to the scene with Maggie, who unfortunately lost the trail at an "Arab well."[17]

Assisting the Jewish Underground Movements

Rudolphina's expertise made her a highly valuable asset not only as an advisor on the use of dogs. After World War II, she advised Zionist underground movements—particularly the Haganah—on how to effectively evade British tracking dogs. Having observed British dog-tracking training techniques, Rudolphina knew their weaknesses all too well.

As World War II ended, the Zionist underground movements refocused their efforts away from assisting the British in fighting against Nazi Germany towards fighting against the British themselves. Because the strict weapon possession regulations of the Arab Revolt (1936–1939) remained mostly in effect after the war, the Zionist underground movements had to stash their weapons and ammunition in secret caches. On June 29, 1946, British authorities in Palestine initiated Operation Agatha, during which the military imposed curfews upon Jewish settlements throughout Palestine while extensive searches and arrests were conducted. In the process, many illegal weapon stashes were uncovered.

Two months later, on August 28, 1946, British military forces initiated Operations Bream and Eel in the Negev region. At dawn, approximately 2,000 soldiers from the 3rd and 8th Parachute Battalions and the 9th Airborne Squadron surrounded Ruhama (established 1911) and Dorot (established 1941), near Gaza, and imposed a curfew. Both settlements were known to be Haganah armed-training centers with large weapon caches. Two days of intensive searches left both settlements in shambles; they uncovered some spent bullet cases and evidence of a shooting range but revealed no sign of a weapon cache. Desperately, the British military called in the dogs of the Corps of Royal Military Police. In Dorot, under a henhouse with a floor of concrete slabs, a crossbred Labrador Retriever soon detected a seven-foot cache that was nearly five feet underground. In Ruhama, a large pile of gravel near a construction site concealed the entrance to a well-disguised stash.[18] From then on, the British began employing mine-detecting dogs in weapon cache searches.

Following these two incidents, the Haganah sought Rudolphina's guidance on how to evade detection. Rudolphina advised the Haganah to refrain from re-digging or fixing ammunition stashes since fresh digging was easily detectable by mine-detecting dogs. That was precisely what she had trained these dogs to do, and though the British may not have been using Rudolphina's dogs against the Haganah, they employed similar methods. Rudolphina counseled the Haganah to dig as many decoy caches as possible to throw off the British dogs. Eventually, thanks to her advice on digging decoys, the British were mostly unsuccessful in searching for weapon stashes.[19]

A far more sensational event involving Rudolphina's evasion advice took place in the Jordan Valley in 1945. A number of Jewish women from local settlements had repeatedly complained of being sexually abused by one Aref Sharida, a local Arab farmer. In one instance, an eighteen-year-old female member of Kibbutz Mesilot was looking for a crossing point of the wadi near the kibbutz. Sharida, who was fishing with a friend nearby, volunteered to assist her. He offered to walk in front of her with his staff to find the shallowest passage. After the woman removed her boots and stockings to cross the water, Sharida assaulted her. The woman cried out for help, but when no one came, she successfully fought him off using her boots. She immediately reported the incident and was examined by a doctor who substantiated her assault allegations.[20]

Though the victim was able to identify her assailant and Sharida was arrested, he was soon released on bail. The authorities' perceived failure to take timely action against Sharida prompted the Haganah to take matters into their own hands: they decided to teach the man a lesson by castrating him. Under the supervision of a local veterinarian, the Haganah members practiced castration on bunnies, using simple household items such as razors and needles to perform the operation.[21]

The six Haganah operatives set out to the man's home one evening, but initially did not find him as he had gone to attend a wedding. They returned after midnight, and under gun threats led the man to a nearby wadi to perform the operation. Unfortunately for both Sharida and for the Haganah members, they lost their bag of ether and cotton balls on their way to the wadi and therefore could not anesthetize the patient. They instead knocked him unconscious and performed the procedure, later leaving his testicles in his hand.[22]

Before embarking on the operation, the six Haganah members consulted Rudolphina on how to evade dog tracking. Rudolphina advised them to flee the scene by walking on the train tracks all the way to Beit Hashita: their prints would be untraceable to the Palestine police dogs due to the various oil scents on the tracks.[25] And indeed, the morning train completely obliterated their scent, and a herd that passed over the tracks further obscured their tracks. On Rudolphina's advice they also took other precautions, walking through water for part of the way, and disposing of their shoes after completing the operation. When the Palestine police-dog unit eventually came to the scene and followed the scent from the bag with the ether, they misled their handlers to Neve Eitan's dining hall, where the dogs could not pick a suspect.

When Sharida was eventually prosecuted for attempted rape, Judge Weldon of the Haifa District Court declined to sentence him to prison time, explaining,

> The accused has had a sufficient punishment by the abhorrent and bestial offence committed on his person by those who have taken the law in to their own hands in such an unjustifiable and uncivilized manner without awaiting the trial and punishment of the accused by the Courts of Justice who are the sole and proper persons to try such offences.[24]

Instead of incarceration, Sharida was placed on an LP 50 bond for good behavior.

Conclusion

Both practically and symbolically, dogs were key to Rudolphina's vision for the Zionist movement. Dogs provided much needed assistance in guarding and protecting newly established Jewish settlements. Her expertise later provided Zionist underground movements with methods necessary to evade detection by the British. Rudolphina was keenly aware not only of the practical support that dogs could provide but also their symbolic value: she was fully conscious of the social and psychological barriers that would have to be overcome to introduce dogs into Jewish settlements as well as the opportunities that dogs presented for closer bonds between nations.

FIVE

Rudolphina Menzel's Contributions to the British War Effort

LEA LEHAVI

During World War II, Rudolphina Menzel played a significant role in the Allied war effort by training mine-detecting dogs for the British army's use on the North African front. The Allied forces faced considerable challenges in North Africa, where the Germans had been laying hundreds of thousands of land mines since 1939.[1] As the fighting intensified there in the early 1940s, the British had already exhausted their canine reserves on the European front and found themselves with a deficit of mine-detecting dogs.[2]

In June 1942, a meeting of high-ranking British officers was held under the auspices of the British Mandate's Veterinary Services Department, which was in charge of all the animals in Palestine.[3] The officers discussed the urgent need for mine-detecting dogs in North Africa and agreed that every effort should be made to train and purchase them as soon as possible. Knowing that Rudolphina was an expert in training dogs to detect mines, they appealed to her for help.[4] Rudolphina in turn consulted with Moshe Sharett, Director of the Jewish Agency's Foreign Affairs Department, to seek his approval before taking any action.

The proposed cooperation between Yishuv leaders and the British was not without its difficulties. On the one hand, as colonial subjects of the mandatory regime in Palestine, leaders of the Yishuv were not eager to help the British. On the other hand, many in the Yishuv wanted to help the Allied forces in their fight against the Nazis. As a compromise, Sharett imposed a condition on the British request: the British must promise not to use the dogs Rudolphina trained for the war effort against Jews in the Yishuv.[5]

With Sharett's blessing, Rudolphina began selling war dogs to the British in the summer of 1942; Sharett granted Rudolphina full authority to manage negotiations on canine matters with the British and deferred completely to her exclusive judgment. The British appointed a special officer to serve as an official liaison between Rudolphina and the British Army, Colonel E. N. Newman Roger.[6] The Colonel subsequently corresponded and met with Rudolphina often to discuss recruitment and training, as did his deputy, the head of the military police in Palestine and Trans-Jordan.[7]

On July 10, 1942, the first cohort of sixty dogs was ready to be transferred from Palestine to the British Army's animal unit, located on the outskirts of Cairo in Almaza. Of these, twenty-two dogs were fully operational; the rest were still in training.[8] By early 1943, dozens of dogs were being transported by the British Army's No. 3 Remount Squadron from Palestine to Almaza on a monthly basis.[9] From there, the dogs were deployed all over the North African front.

From April 2, 1943 until the end of the war, weekly reports on the number of dogs purchased by the British and their conditions were received by the Veterinary Department of the Mandate government. An April 1943 report noted that twenty war dogs were purchased in Haifa and were described as "in good condition."[10] Though Rudolphina's name is not mentioned, it is clear from the description that they were purchased from her kennel in Haifa. A report from May 1943 recorded the purchase of thirty-nine dogs and mentioned that the availability of high-quality Boxers remained good and that "breeding is carried out more extensively now... it is anticipated that there will be no difficulty in purchasing dogs in the future in large numbers."[11] A December 1943 entry noted that "four dogs—three Boxers and one Alsatian—were purchased from Dr. R. Menzel Kiryat Motzkin."[12] Rudolphina's name appears in all subsequent reports on dog purchases.[13] Though the war diaries mention occasional single dog purchases from individual private breeders, those were unusual.

Rudolphina's institute faced dire financial challenges during World War II, and she desperately needed money to care for and train her dogs. The relationship Rudolphina established with the British was therefore mutually beneficial: they benefitted from her highly trained dogs, and her institute benefitted from the income generated by their sale.

The British Army ultimately deployed over four hundred mine-detecting dogs trained by Rudolphina or according to her methods, and the British war diaries recorded British soldiers of all ranks praising the heroism and excellence of these dogs.[14] One of the officers reportedly remarked that the only disadvantage of the dogs was the fact that they couldn't get more of them.[15] Articles in the *Palestine Post* also extolled the virtues of these dogs, which contributed greatly to the sense of pride amongst Jews in the Yishuv for being able to contribute to the fight against the Nazis.[16] Despite Rudolphina's great success training mine-detecting dogs for the British during World War II, her authority as the supreme expert on military dogs in Palestine was challenged not long after the war ended. In November 1947, the Haganah began to lay the groundwork for an official canine unit in anticipation of its eventual integration into the established army of the new state.[17] At that time, it seemed inevitable that Rudolphina would be put in charge of the unit. In the 1947 annual report for her institute, she wrote:

> In November we addressed the Haganah Chief of Staff Yaakov Dori and requested to expand the Institute's reach of activities to comply with future demands. Yaakov Dori decided that Dr. Menzel would be the sole leader and professional address for all canine issues. Moshe Dayan was sent to us and together we started planning the canine activities and organizational structure for the soon-to-be-formed Army.[18]

Consequently, Rudolphina assumed a commanding tone when she wrote an article for the dog trainers and handlers of Palestine as preparations for war accelerated:

> Not every generation is as blessed as ours, for we are lucky to witness the dreams of our youth come true...a hope of two thousand years is coming true...We, the Jewish dog handlers, will see how our own dogs will take their place in the building of our homeland, as they have so far taken their place as our loyal companions. But alas, we do not have time to celebrate, as this is the hour of preparation. Each of us at our post.[19]

Unbeknownst to her, this would be the last time Rudolphina addressed her handlers and trainers as leader and mentor. In February 1948, Yaakov Dori, Chief of Staff for the Haganah, sent her a letter informing her of the Haganah's revised plans for a canine military unit:

> We have decided to found a "Central Military Dog Training Camp" for all of our military unit's needs. We ask that you be the professional supervisor. We assume these supervising activities will keep you busy for half a day's work, every day...We hope that with the foundation of the Hebrew State, the Institution will be fully built and your dream and hard work will bear its fruit. As for this "Central Training Camp," we could not achieve it if not for your hard work and devotion.[20]

This letter was bittersweet. It informed Rudolphina that she would no longer be in charge of the military canine unit as the Yishuv transitioned to statehood; instead, she would only oversee the educational-professional aspects of the canine issues as a part-time job. While acknowledging her significant contributions to establishing the entire foundation of dog-training in the country, Dori basically sidelined her. In her place, one of Rudolphina's first students from her 1937 dog-training course, Abraham Zirlin, was assigned to be the commander of the new canine unit.[21] Zirlin's military expertise as a soldier, as well as his experience as a dog handler for the Haganah, made him the perfect candidate for the job. Zirlin was put in charge of the canine unit's management and administration, all things Rudolphina used to supervise, and he was also put in charge of Rudolphina herself.[22] It seemed that this move was well-planned by the new Israel Defense Force (IDF) command, as illustrated by the letters of complaint Rudolphina wrote to Moshe Dayan.[23] Rudolphina had been replaced.[24]

Rudolphina's demotion to part-time educational supervisor in 1948 understandably resulted in a rift between her and the IDF. She retained an honorary title that enabled her to maintain occasional work with military canines and was appointed the Government Advisor for dog breeding and training, in which position she served as a member of a formal, government interministerial committee that addressed the use of guard dogs for security needs.[25] The committee was led by her old Haganah associate Yaakov Pat. This role was primarily an honorary one, given by the new Israeli establishment to thank her for her years of contribution. Thus, when the Israel Defense Force was officially founded in June 1948, Dr. Menzel's military career came to an end.

In June 1949, the IDF headquarters decided to transfer the canine unit—which was then under the command of the IDF's animal services—to the military police, as was common in the armed forces of other nations.[26] The IDF's canine unit was active until 1954, when it was closed for twenty years; it was refounded in 1974 as a special operations combat unit under the command of the infantry corps' headquarters.[27] The IDF's current elite canine unit "Oketz," considered to be amongst the world's best, continues to use dog-training methods pioneered by Rudolphina. The link she created between Zionism and dog training was essential not only to the Allied war effort during World War II but also to the establishment of the state.

SIX
Personal Recollections of Rudolphina Menzel and Her Canaan Dog Breed

MYRNA SHIBOLETH

Rudolphina Menzel was a tiny woman with an enormous presence. I first met her in 1970, when she was seventy-nine, but "old" was never a word that could be attached to her. Even at this age, she was a ball of energy, running from one task to another with a mind as sharp as it ever had been and a fierce sense of humor. Her love for dogs was clear when we met at her beloved Institute for the Orientation and Mobility of the Blind, as was her love for people and her desire to do the best for them. Meeting her and having the chance to learn from her was life changing for me.

As a young and very naïve and inexperienced dog person, I was in awe of this woman who was so well-known and respected in the field. But she was not at all standoffish. She was very vivacious, full of energy, able to converse fluently in a number of languages, and able to laugh at herself. Her personal house dogs were Boxers and were quite undisciplined; she laughed as she told us, "I can tell you how to train dogs, I just can't do it with my own dogs!" She expected those who worked with her to be as dedicated as she was, and her standards for the performance of all tasks were very high, but she had the total respect of everyone. At times considered controlling and temperamental, she was a perfectionist who wanted things done her way. The dogs she trained, both for military purposes and as guide dogs for the blind, had to perform in life and death situations, which justified her strictness and unwillingness to relax her criteria. She had many interests and projects in addition to dogs. For example, she wanted to gain recognition for the feral cats found all over the country; in her opinion, these cats, with tufts on the tips of their ears, were direct descendants of the original Egyptian cats, and she felt they should also be preserved.

The Canaan Dog

One of the topics that most occupied her attention, both as a scientist and as a cynologist, was the local dogs native to Palestine. She felt that these dogs were as nature had created them and had not been changed by any form of selective breeding other than the selection of nature for survival; as a result, they were perfectly adapted to living in the harsh environment and capable of subsisting on the meager resources available. The area where she lived near Haifa was quite open, with very few buildings, and was surrounded by dunes. She could observe these dogs living free in the area and following the Bedouin flocks on their annual grazing route. Because she considered these dogs to be the link between the common household breeds and their wild canine ancestors, she believed they should be protected and preserved. She was concerned they were becoming endangered in their natural habitat by mixing with the domestic breeds that were increasingly found all over the country. In addition, she knew that the rabies control programs managed by the government threatened to destroy these packs of free-living dogs.

With the Bedouin, these dogs lived in a kind of partnership, providing services of guarding, alerting, and driving away predators; in exchange, they were given some food and the possibility of gaining a certain amount of safety by remaining in the vicinity of the Bedouin camp. Even the dogs that were not working dogs and that "belonged" to no one tended to stay in the general vicinity of human habitation. Though they were not pets and did not look for human companionship or friendship, Rudolphina believed they were very amenable to developing a relationship with humans. She gradually became quite amazed by the ease with which these dogs built up a close relationship with people and how this friendship with humans changed them from shy, suspicious, wild animals to devoted dogs that were happy to be pets or to learn a task.

Rudolphina decided to collect these dogs and try them out as working dogs. She obtained dogs from the Bedouin and from other locals who became aware of her interest in them. People brought her dogs from various sources; some had been free-living dogs in various parts of the country or had been born to these small packs of free-living dogs and adopted as pets, and some had been pets and guard dogs.

Recollections of Rudolphina and Her Canaans

To develop the Canaan dog breed, Rudolphina carefully examined the phenotypes of these dogs as well as their general temperaments. She possessed very definite ideas of how these dogs should look and act, based both on her own experiences with the breed and observations of them with the Bedouin. She also relied on the reports of writers and travelers who had observed the fauna in Palestine and written descriptions of these dogs in the past.

Everything about these dogs had to be what she called "natural": there could be no exaggerated features that interfered with their ability to function. To her, the ideal Canaan dog was a medium type of pariah dog, part of a continuum of pariahs found throughout the Middle East, all related and descended from the original natural dog of the region with only minor physical differences according to the locale they were found in.

One description that she wrote to explain her ideas about pariah dogs was as follows:

> 1. Not every stray dog in the Orient is a pariah. We also find stray dogs in countries with an advanced village and urban culture, but they can usually only live for a short time.
> 2. Not every pariah is masterless. Pariahs as herd dogs and village guards have been around for thousands of years across the Orient and beyond.
> 3. Not every pariah lives in the streets of cities. This sentence rightly turns against the so often used term "street dogs."
> 4. Not every oriental street dog is a crossbreed; i.e. there are street dogs that could be preserved in their territory as an unmixed form.[1]

To establish a Canaan dog breeding program, she began to selectively combine the best examples of these pariah dogs. She raised dozens of litters of puppies and kept her own stud book, which contained every entry covering the early years of the breed.

She very much believed in the importance of the research and study of these dogs and of their value to science. There was a great danger of the pariahs disappearing due to crossbreeding, and she felt, along with other zoologists and scientists that she was in contact with, that it was essential to devote research to them at this time. In 1960, she wrote her monograph *Pariahunde* about the breed, which described her research and experience with the Canaan dogs and her objectives for the breed. This booklet remains relevant today. The first breed standard was included in this book, which became the basis of the standard subsequently accepted by the Federation Cynologique Internationale in 1966. One of my most precious possessions is a copy of her *Pariahunde* that she presented to me with the inscription, "To a pioneer in the breeding of Canaans."

Rudolphina very much believed that to preserve this breed and gain recognition for it, there would have to be international interest. She was very interested in making contact with people abroad and was happy to send top-quality dogs as examples of the breed to potential breeders in other countries. In 1965, she sent four dogs to Ursula Berkowitz in California, who was interested in breeding Canaan dogs; subsequently, Rudolphina sent a number of additional dogs to the United States. She also sent a dog to an old friend of hers in Germany, Freia Eisner. This dog was trained to assist Mrs. Eisner, who was disabled. She was proud to have an Israeli dog and was well-known in Berlin for walking around with him. In the late 1960s, Rudolphina sent the first registered Canaan dog to England.

In correspondence that began in 1968 with Connie Higgins, one of the first Canaan dog breeders in England, Rudolphina described many qualities that she considered important and essential to the breed.[2] Rudolphina sent Higgins a male Canaan dog named *Tiron me B'nei Habitachon* to use as a stud for Connie's Syrian-born bitch (*B'nei Habitachon* was Rudolphina's kennel name, meaning "Children of Security"). She described the ideal puppy Connie should endeavor to breed as being square with an especially short back, a characteristic that Rudolphina valued since compact dogs functioned most efficiently in the local climate conditions. She described the ideal ears on a spectrum from button ears to semi-prick to full-prick, adding that full-prick ears were "the form of the future" and that she preferred them. The puppy's preferred coat type was harsh and strong, short to medium in length, and with a profuse undercoat. This type of coat provided protection from all types of weather, thorny desert undergrowth, and parasites. Additionally, the very thick undercoat, especially on the neck, was an excellent defense against attacks by predators. As for colors, Rudolphina wrote that she personally preferred black or red-brown dogs with large white markings, but any colors with the exceptions of brindle and gray, common to shepherd dogs, were allowed; those colors, if found, were rejected for registration and breeding, as her opinion was that they were the result of impure descent. Above all, she preferred black and white coloring, which she considered most typical.

Temperament was of great importance to Rudolphina. She believed there was a genetic component of behavior traits, which was a very cutting-edge idea at this time. These dogs were expected to be watchful, alert to something approaching from a distance, and cautious with strangers.

Recollections of Rudolphina and Her Canaans

Rudolphina was not interested in turning these natural dogs into another generic breed. The characteristics that enabled them to survive in their natural habitat needed to be preserved. She felt that "the individual pariah dog is ready at any time to take up an appropriate position on the line between the wild state and domesticity and will adapt his behavior toward man to correspond to the situation."[3]

Rudolphina described the method of verifying that puppies conceived from an unknown parent were pure Canaans as the *miun* (selection). In order to be registered in the Canaan dog stud book, dogs with unknown parentage had to be mated with a dog of known origin with an established pedigree. Rudolphina subsequently wrote to the English Kennel Club to request their acceptance of the registration of the mother and her expected pup:

> I am in contact with Mrs. Connie Higgins of West Malvern, owner of the Canaan Dog Shebaba. By means of a number of pictures, made in various positions, which Mrs. Higgins has sent to me, I have reached the firm conclusion that Shebaba is a typical Canaan bitch in every respect, representing the best collie-like form of that breed. I would give her the classification of, at least, "very good" at any show in this country. We would be inclined to register her in the Stud Book of our kennel club, Canaan Dog Section, in order to make breeding of these dogs possible in England. The bitch is, according to the photos, perfectly typical. Unfortunately, a pedigree cannot always be proved with a natural race. Most Canaanites with whom we ourselves have established that breed have been, of course, without pedigree, or we have known at most father and mother. We are interested that breeding would be undertaken with Shebaba. Therefore, we have offered Mrs. Higgins to send her, as a present, a stud dog for her bitch. We have chosen a typical Canaanite who, of course, is registered with our kennel club, and we are about to forward him to her.
>
> After getting pictures of the first trial litter I intend to ask you, dear Sir, to require an expert opinion on the puppies...and, on the strength of that, to register them. I realize that such a registration would be very uncommon in England, a country in which dog breeding is on such a high grade. But there is no other way left if you are interested in adding a new natural breed to your "cultural races." We ourselves had to do the same with the Canaanite breeding in our country. England has always been interested in transforming natural races into cultural breeds. It was your fellow countrymen who have succeeded in breeding the pure Basenji. Please help us in the same way with our Canaanites.[4]

In the early 1970s, we began our work to preserve and breed Canaans at our kennel in Shaar Hagai, located just off the road between Tel

Aviv and Jerusalem. In order to establish our operations, we were very dependent on Rudolphina's knowledge and expertise. I frequently drove to Haifa to bring dogs for her examination and relied on her assessment as to whether they were pure Canaans; she also advised on whether or not to do test breeding. She had a very keen eye and gave me very comprehensive and clear explanations of her decisions about what was typical and worth preserving versus what was not and why. She recognized every small sign that a dog was not pure. Despite the long drive—in this period, about three hours each direction—I tremendously enjoyed going to see her and with each visit learned new things. Her knowledge about dogs was encyclopedic, and she was excellent at imparting it.

We were lucky enough to be given a number of dogs from her kennel to become the foundation of our Canaan dog kennel, some from several generations of her planned breeding, and some directly from free-living or Bedouin packs. These dogs included *Laish me B'nei Habitachon* who became our foundation stud dog and is an ancestor of many of the Canaans present today. He was a wonderful dog from one of the last litters she bred before retiring, and she considered him to be an ideal example of what a Canaan should be, one of the best she had ever bred. He remained a symbol for us of what to breed for. Laish, known at home as Simi, became the first Israel Champion in the breed and was among the first to bring the breed into the public eye.

In addition to dogs from Rudolphina's kennel, we began to collect dogs from the Bedouin and, whenever possible, from free-living packs. We also were contacted by people who had pets at home that they thought might be Canaans. Rudolphina was very pleased that we were continuing to breed Canaans. The preservation of this original breed was very important to her; she felt that it was important to cynology to have breeds of the original dog for us to learn about natural behaviors, instincts, drives, relationships, and so on. Other breeds had, over the years, been changed so much that it was only possible to learn from the natural breeds what a dog really was and what its characteristics and behaviors had been when it first decided to become a partner to man. From them, we were able to learn more about understanding and building successful relationships with dogs. Rudolphina was at the forefront of scientific thinking about these matters.

I was fortunate enough to have attended the last judge's training course that she gave in 1971 for the kennel club. It was held over a weekend at her institute in Kiriat Hayim and was attended by prospective student

Recollections of Rudolphina and Her Canaans

judges sent by their breed clubs. The program consisted of several days of lectures given by experts on canine structure and movement, breed specific details, interpretation of breed standards, behaviors, and more. A number of the lectures were given by Rudolphina herself, and despite her advanced age, she was an excellent lecturer, clear and precise, and presented the relevant information spiced with humor and anecdotes. It was an unforgettable experience.

The project that was perhaps the closest to her heart was the training of guide dogs for the blind. The first dog she trained for guiding was a Boxer (Rudolphina loved the breed and used them for many of her projects), but she was thrilled when she was given a purebred Labrador from overseas to use as a stud dog from one of the foremost schools for the blind. She used him to build up a breeding colony of Labradors in Israel that were selected and bred specifically for guiding.

Rudolphina was very strict in her criteria both for the dogs intended for guide-dog work and for the blind persons who were to receive them. She developed a test to analyze the abilities of the blind person to learn mobility techniques and to assess whether they were a suitable candidate for a guide dog. She also had very particular expectations for the dog's behavior and analyzed it throughout the training period. She developed methods of training that were quite innovative. One example is a cart that she developed, as tall as a normal person, that was hitched to the dog. The dog had to pass low obstacles that could be passed by the dog alone, but with the cart, the dog's progress was stopped and it could not move forward. Through training, the dog learned how to avoid the obstacle to be able to continue forward just as it would have to do with a blind person.

Rudolphina's comment on this training was: "The Bible commands us not to place an obstacle before the blind. Our slogan is basically the opposite: Place an obstacle in the way of the blind, so they will learn to overcome it."

Rudolphina was honored by the Jewish National Fund (JNF) in 1966, and her name was inscribed in the Golden Book of the JNF for her distinguished contributions to the Zionist movement. She never lost her passion for Zionism, her optimism for the future, or her love and support for the young people that she taught and encouraged. She believed that cynology could be developed to a high level in Israel and that the Canaan dog could gain international recognition as an accepted and valued breed. She has, in the last years, been proven right. Without her efforts, we could not have reached these goals.

SEVEN

Rudolphina Menzel in Israeli Culture and Historiography

TAMMY BAR-JOSEPH

GIVEN THE MAGNITUDE of Rudolphina Menzel's contributions to the protection of the Yishuv, her important military role during the War of Independence, her pioneering work with the blind in Israel, and her international reputation as one of the foremost cynologists of her day, why are her accomplishments not more well-known in Israel, either in the popular imagination or in Israeli historiography? To date, only two unpublished theses have been written about Rudolphina Menzel in Hebrew, and only a few mainstream features have appeared in Israeli newspapers.[1] The scattered references to her in various Hebrew-language encyclopedias, magazines, newspapers, and websites are few and far between.

Rudolphina's virtual disappearance from the contemporary national memory in Israel is made more curious by the fact that she was a well-known public figure during her lifetime. Articles by and about her dog training often appeared in the popular Hebrew-language press. Dozens of articles by and about her appeared in *Davar* in the 1940s. The opening of her guide-dog institute was featured in a 1953 article in *Haaretz*.[2] Coverage of her expert testimony based on her experience with tracking dogs appeared in the *Herut* newspaper in 1965,[3] and a review of the Menzels' children book about dogs and cats appeared in *Maariv* in 1969.[4]

The first and only Hebrew-language encyclopedia reference to Rudolphina Menzel appears in the *Encyclopedia of the Pioneers of the Yishuv and its Builders* published in 1947. The entry on Rudolphina is not comprehensive and includes few details about her biography, her areas of expertise, and her professional status:

In 1928 Menzel brought with her husband the proof that each person has an individual scent (special to him). The Menzel couple worked together with the German and Austrian police and military, lecturing at several international scientific congresses on issues related to dog psychology and working with dogs. In 1935 they presented at a congress held in Frankfurt the first results from their study on the heredity of behavioral traits of dogs based on their research of 7 generations of Boxers. In 1934, Menzel visited Israel at the invitation of Yaakov Pat, one of the commanders of the Haganah... in 1937 they were the main lecturers on dog psychology at the International Congress of Cynologists held in Paris. That same year, an institute for the study and training of dogs was established in Kiryat Motzkin in Haifa at the request of the Haganah. During the Second World War, at the request of Moshe Sharett, many dogs were prepared for the war effort of the Allied armies abroad.[5]

Over three decades later in September 1973, shortly after her death, the entire issue of the Hebrew-language dog magazine *Hakelev* was devoted to tributes of Rudolphina written by friends and associates, including:

"She Was an Important Person" by Hans Raba from Switzerland
"Doctor Menzel: An Evaluation by a Student" by Rafael Freedong
"My Years with Dr. Menzel" by Ephraim Benhar
"The Way of Dr. Menzel" by Abraham Tzur[6]

In 1987, an additional tribute appeared on the website of the Israeli Kennel Club written by her former assistant Ester Cohn, though it was subsequently removed.[7] In 2006, Ester and her husband Alex published a more formal article in the Hebrew-language journal *Animals and Society* about working with the Menzels in the 1940s.[8] Here, the Cohns describe, among other things, how Ester first met Rudolphina when she came to her girl scout troop to do a dog demonstration in order to recruit volunteers for her canine institute.

Rudolphina's presence on contemporary Hebrew-language websites is negligible. In 2010, an article posted on the Israel Defense Force Archive's website featured an online exhibition of the canine corps in the Israeli army and included a short article about Rudolphina's role training dogs for the Haganah.[9] In 2013, the Central Zionist Archives posted a short biography of Menzel called "Rudolphina's Dogs" on their website. The post provided a brief overview of her life and mentioned that Menzel trained police dogs in Austria to follow commands in Hebrew, noting "with an ironic sense of humor reserved only for history itself, the first dogs of the German army, which would later become the Nazi army, obeyed instructions in Hebrew only."[10] The website for the Israeli Center

Rudolphina in Israeli Culture and Historiography 133

for Guide Dogs makes mention of the guide-dog institute she established in 1952: "There had been a woman who trained guide dogs in the 1960s named Dr. Rudolphina Menzel; however, when she passed away, the program was abandoned."[11] Finally, there is a short entry in Hebrew about Rudolphina on Wikipedia, which provides a general overview of her life as recounted in the previously mentioned sources.[12]

Rudolphina's disappearance from the Israeli popular consciousness is compounded by her invisibility in Israeli historiography. Her gender may have contributed to this absence, given traditional academic biases against centering women in history.[13] Yet in the past few decades, feminist historians have worked tirelessly to restore women to the Israeli national narrative and to memorialize their contributions to Israeli institution-building, politics, and culture.[14] Certainly Rudolphina's gender doesn't explain why she remained invisible even to them.

Rudolphina's area of expertise may be part of the reason she has been overlooked by Israeli historians. According to a recent study, there are few intellectual biographies of Israeli scientists in Hebrew; most of the biographies written in Hebrew to date focus primarily on political figures: Zionist leaders, Israeli statesmen, and military heroes.[15]

Her age and her country of origin may be additional reasons Israeli historians have overlooked her. Rudolphina immigrated to Palestine at the age of forty-seven and made all her contributions in her later years, confounding common myths that the country was built by young pioneers. Perhaps she has been overlooked by Israeli historians because they were unduly influenced by popular cultural biases that positioned German-speaking Jews as bourgeois, elitist, and culturally foreign. But these are also unsatisfactory explanations for Rudolphina's absence from the national narrative. The lives of many relatively older Israelis, let alone many notable German-Jewish immigrants from the 1920s and 1930s, have been copiously documented and their contributions memorialized by Israeli historians.

While unlikely, perhaps Menzel's professional contributions to the state's defensive apparatus have been deliberately concealed because the role military dogs play in Israeli national security was classified as top secret. Or perhaps Rudolphina's association with the Canaan dog contributed to her marginalization. Ambivalence about dogs in general, and the Canaan dog in particular, has deep roots in Israeli society. The cultural challenges facing the Canaan dog were threefold: the traditional Jewish ambivalence to dogs made Israeli Jews unlikely enthusiasts for a national dog breed; the breed's generally skittish disposition often made it difficult to train and enjoy; and its association with the Bedouin played into a general—if selective—antipathy towards Arab cultural objects, even more so because the Canaan dog holds little symbolic or cultural value for the Bedouins themselves (unlike salukis, horses, or camels).[16]

While her gender, profession, age, country of origin, and association with the Canaan dog may have something to do with Rudolphina's relative invisibility in Israeli national memory, these oversights, absences, and omissions are better explained by the fact that she did not fit neatly into any recognizable Israeli social categories. She was not a mother, she was not a farmer, she was not a politician, she was not a soldier, and she was not a war hero. If she had sacrificed her life for the Yishuv in the war with the Nazis or during the establishment of the state and the Israeli wars, she might have become a commemorated heroine like the partisan Sarah Aaronson or the paratrooper Hannah Szenes.

Moreover, Rudolphina did not occupy any of the professions traditionally associated with women: she was not a nurse, a social worker, or a teacher in the traditional sense. She resisted easy categorization in other ways as well: she was an urban, cosmopolitan woman who trained dogs to work in fields, forests, and mountains. She worked with the blind at a time in Israeli history when cultural sensitivity and understanding of the disabled was limited. Moreover, she confounded traditional gender expectations in at least two ways: by being a woman who wielded power over men as an acknowledged expert and by maintaining a strikingly egalitarian relationship with her husband.

I suggest that the traditional categories of Israeli historiography need to be stretched to appreciate Rudolphina's unique contributions to Israeli society. Although she was not a mother, she was a nurturer whose "children" were dogs who contributed to national wellbeing. Although she was not a farmer, she understood the textures and smells of Israeli soil as part of her work with scent hounds. Although she was not a politician, she was a highly persuasive campaigner who mobilized thousands to realize her vision. Although she was not a soldier, she trained dogs to be soldiers, who in turn played key roles in providing security for Jewish lives and settlements. Although she was not a "teacher," she taught hundreds of dog handlers how to train and breed dogs, thereby laying the foundation for successive generations of expert dog training. Although she was not a nurse, her efforts to organize public health officials and veterinarians to eradicate rabies and other canine diseases played an

Rudolphina in Israeli Culture and Historiography 135

essential role in improving overall public health conditions in the country. The canine infrastructure she established has evolved into a diverse and robust network of Israeli veterinary services, animal shelters, dog parks, dog beaches, and dog-training programs.

Much research remains to be done on Rudolphina's remarkable contributions to Israeli culture, politics, and society—particularly with regard to her foundational work in creating the Israel Defense Force's elite canine corps Oketz and her pioneering work in providing support for the sightless. There are some indications that greater Israeli interest in her may be emerging. Rudolphina's life story was included in *True Legends: Fifty Women to Grow Up with in Israel*, a 2019 Hebrew-language children's book about Israeli national heroines.[17] In it, an illustration appears of Rudolphina giving commands to three bouncing dogs—a Canaan dog, a Boxer, and a Pointer—and is captioned with a quote she and Rudolph wrote in their final children's book: "Friendship with the dog is the shortest and most convenient way for a person who wants to get closer to the animal world."[18]

Acknowledgments

IT IS DIFFICULT to express the enormity of my gratitude to all the people who helped and supported me throughout this project. First, thank you to Tslila Zagagi for inspirational dog walks in Beer Sheva many years ago. I am deeply indebted to Gail Goodman, Tzviah Idan, Zafra Sirik, Jill Terry, Allan Degen, Shaher Elmekawi, and Juma'a for generously sharing their expansive knowledge about dogs during an early iteration of this project. Carmel Shalev, Elly Teman, and Tali Berner helped enormously with logistics and sources at that time. Meeting Myrna Shiboleth at Shaar Hagai Kennels while she still lived and worked there provided an invaluable introduction to the Canaan dog.

Susannah Heschel, Harriet Ritvo, Chris Pearson, Helen Epstein, Lilian Handlin, Naomi Seidman, Susan Miller, and Matti Bunzl offered helpful comments on early drafts of this manuscript and could not have been more encouraging or supportive. Aviva Dautch provided exceptional suggestions, wise corrections, and thoughtful edits to the penultimate draft, for which I am deeply grateful. My brother Jon Kahn added his own inimitable insight and commentary throughout the writing process.

In addition to their extraordinary warmth, Irit, Israel, and Dor Aharony provided critical help with Hebrew translations. Sharon Bar Shaul and Ilana Szobel patiently helped translate old mimeographed Hebrew letters and articles with me over Zoom. Alex Dennett kindly produced a near-complete translation of an important and complicated French scientific article. Barbara Schmutzler, Susan Hecker Ray, Ilse Browner, and Annkatrine Gates were unbelievably helpful with German translations.

Xenia Cherkaev, Gili Stolarski, Nimrod Samoray Levi, Dennis Allon, Dafna Shir-Vertesh, Joseph Terkel, Hiali Gross, and Ester Cohn provided crucial help gathering sources and sharing personal perspectives. A particular thank you to Ester for sharing her singular and important recollections.

George Waltuch (Rudolphina's nephew) and Serena Fox (Rudolphina's grandniece) kindly provided wonderful family reminiscences. I am

Acknowledgments

especially grateful to Serena for generously sharing her personal archive of photographs, letters, and other mementos of the Waltuch family.

For archival help, I am enormously indebted to the staff at the Central Zionist Archives, the Leo Baeck Institute, and the Haganah Archives. Violet Radnofsky was very helpful with sources at Widener Library. Margaret Doherty's excellent suggestions about best practices in biographical writing were invaluable. Noah Feldman—and the coronavirus—accorded me time and space to work on this project. I consider Noah's generosity of spirit and enthusiastic support an absolute gift.

The contributors to this volume, Monika Baár, Rachel Koriat, Binyamin Blum, Lea Lehavi, Myrna Shiboleth, and Tammy Bar-Joseph were exceptionally patient and forthcoming with their time and expertise. This book would never have seen the light of day had it not been for Sylvia Fuks Fried's advice and encouragement.

For their kindness and constancy while I worked on this, my deep thanks to Angela DeVecchi, Lydia Vagts, Deb Neiman, Roni Breite, Thea Breite, Susan Moser, z"l, Annie Valva, Beatriz López-Flores, Adam Weisberg, Jessie Meltsner, Susan Harris, Chuck Freilich, Elizabeth Marks, Maribeth Kaptchuk, and everyone in the Bolton-Fasman and Kahn-Taussig-Varhelyi families. And finally, thank you to Esther and Solomon for their excellent company.

Notes

Notes to the Preface

1. The exact number of dogs that served in the 1948 war remains unclear; estimates range between 300 and 2,500, the former reported in a personal interview with Ester Cohn (March 2020), who served in the Haganah's canine corps, while the latter appeared in a newspaper article written by Rudolphina's niece Anita Fox. Anita Fox, "Land's Best Friend," *MetroWest Jewish News*, May 1, 2008, A54–A55.

2. "Hit'aḥadut ha-Yiśre'elit le-khalbanut," *Hakelev, Journal of the Israeli Kennel Club*, no. 9 (September 1973).

3. Interview with Ester Cohn (March 2020): "In general, Dolphi was the one who had the flashes of inspiration and ideas; the impulsive, purposeful woman was the engine, so working with her was not always easy. Rudi arranged her data slowly and carefully and formulated what was written logically. He had a better way of dealing with people and balanced things out. Since he had to pursue his extensive practice as a doctor, the practical part—the kennel operation, the recording of observations—fell primarily to Dolphi."

4. Handwritten notes in the pages evaluating Rudolphina's essay indicate she was paid the sum of $125 on February 16, 1941. Gordon Allport, Sidney Fay, and Edward Hartshorne, "My Life in Germany Contest Papers, 1940" (unpublished manuscript, 1940), printed volume.

5. Harry Liebersohn and Dorothee Schneider, "'My Life in Germany before and after January 30, 1933': A Guide to a Manuscript Collection at Houghton Library, Harvard University," *Transactions of the American Philosophical Society* 91, no. 3 (2001): 87.

6. Verena Wagner, *Jüdisches Leben in Linz 1849–1943* (Linz: Wagner Verlag, 2008). Detlef Garz, "Rudolphine Menzel," in *"Wie ein Schatten ging ich meinen Weg zu Ende": Emigrantinnen aus Wissenschaft und Kunst* (Budrich: UniPress, 2014), 175–202, www.linzwiki.at/wiki/Menzelweg/, accessed December 1, 2021.

One · The Extraordinary Life of Rudolphina Menzel

1. Gordon Allport, Sidney Fay, and Edward Hartshorne, "My Life in Germany Contest Papers, 1940" (unpublished manuscript, 1940), printed volume, 2.

2. Michael Berkowitz, *Zionist Culture and West European Jewry before the First World War* (Cambridge: Cambridge University Press, 1993).

3. John W. Boyer, *Culture and Political Crisis in Vienna: Christian Socialism in Power, 1897–1918* (Chicago: University of Chicago Press, 1995), 177.

4. Allport, Fay, and Hartshorne, "My Life in Germany Contest Papers," 22.

5. Allport, Fay, and Hartshorne, "My Life in Germany Contest Papers," 29.

6. According to Rudolphina, Frau Glockel later recalled most of Rudolphina's responses were uttered "in a defiant, aggressive tone" and that she was "always ready to oppose." Ibid., 29.

7. Ibid., 29.

8. Coincidentally, one of the Glockels' neighbors was Hermine Santrouschitz, later known as Miep Gies, who helped Anne Frank in Amsterdam and saved her diary.

9. Three mass political parties emerged in the turbulence of post-Hapsburg Vienna. Viktor Adler's Social Democrats appealed to the polyglot, assimilated, urban proletariat; Karl Lueger's Christian Social Party to the antisemitic masses of Germanized petit-bourgeoisie; and Georg Ritter von Schonerer's Pan-German Nationalists to the rest of the middle classes and anyone else who was anticlerical, antisocialist, and antisemitic. For a politically-attuned young woman like Rudolphina, the attraction of the Austrian Social Democrat's ideology, let alone her affection for the Glockels, logically attracted her to the party.

10. "Women on the Move 1848–1938," Austrian National Library, accessed February 15, 2022, https://fraueninbewegung.onb.ac.at/node/869.

11. Allport, Fay, and Hartshorne, "My Life in Germany Contest Papers," 31.

12. Ibid., 88.

13. Ibid., 58.

14. Ibid., 61.

15. Rudolphina's dissertation was entitled "Über die Charakterisierung der Harnpentose" [On the Characterisation of Arabinose], noted in Detlef Garz, "Rudolphine Menzel," In *"Wie ein Schatten ging ich meinen Weg zu Ende"—Emigrantinnen aus Wissenschaft und Kunst* (Budrich: UniPress, 2014), 184. And Rudolph Werner Soukup and Robert Rosner, "Scientific Contributions of the First Female Chemists at the University of Vienna Mirrored in Publications in Chemical Monthly 1902–1919," *Monatshefte für Chemie* [Chemical Monthly] (2019): 969, https://doi.org/10.1007/s00706-019-02408-4.

16. Garz, "Rudolphine Menzel," 186.

17. Allport, Fay, and Hartshorne, "My Life in Germany Contest Papers," 68.

18. Michael John, "Jews as Consumers and Providers in Provincial Towns: The Example of Linz and Salzburg, 1900–1938," in *Longing, Belonging, and the Making of Jewish Consumer Culture* (Leiden: Brill, 2010).

19. This was a decision for which Rudolph paid a high price; he was later accused of abusing his official authority and had to hire a lawyer to clear his name. Allport, Fay, and Hartshorne, "My Life in Germany Contest Papers," 74.

20. Verena Wagner, *Jüdisches Leben in Linz: 1849–1943* (Linz: Verlag, 2008), 1076.

21. Allport, Fay, and Hartshorne, "My Life in Germany Contest Papers," 97.

22. Wagner, *Jüdisches Leben in Linz*, 1052–58.

23. Allport, Fay, and Hartshorne, "My Life in Germany Contest Papers," 104.

24. Joseph Bodingbauer, *Wesensanalyse für Junghunde* [Illustr.] (Vienna: Österreichischer Kynologenverband 1969).

25. The seemingly whimsical and nonsensical name Pazzos-Naxos rhymes. "Pazzos" means crazy in Italian, and "Naxos" refers to the Greek island.

26. Bodingbauer, *Wesensanalyse für Junghunde*.

27. Konrad Most went on to invent methods for training German military dogs in World War I to carry messages, patrol territory, and pursue and attack perceived enemies. Because these dogs were expected to perform specific tasks in dangerous situations, Most believed their success depended on complete obedience to their master's commands.

28. Phyllis Poduschka-Aigner, "Menzel Rudolphine und Rudolph" in *100 Jahre Österreichischer Kynologenverband, 1909–2009* (Vienna: Österreichischer Kynologenverband, 2009).

29. Thomas Mann, *Bashan and I* (Philadelphia: University of Pennsylvania Press, 2003), 97.

30. Allport, Fay, and Hartshorne, "My Life in Germany Contest Papers," 104.

31. Ibid., 173.

32. Rudolph Menzel, "Hounds of Hope: The Strange Story of Dog Breeding in Palestine," *Canadian Jewish Chronicle*, 1935: 8–9.

33. Rudolph Menzel and Rudolphina Menzel, *Welpe und Umwelt* (Leipzig: Schöps, 1937).

34. Ibid., 2.

35. Adam Miklósi, *Dog Behaviour, Evolution, and Cognition* (New York: Oxford University Press, 2015), 299.

36. Sherman Ross, "Some Observations on the Lair Dwelling Behavior of Dogs," *Behaviour* 2, no. 3 (1950): 144–62, www.jstor.org/stable/4532701.

37. J. P. Scott and Mary-Vesta Marston, "Critical Periods Affecting the Development of Normal and Mal-Adjustive Social Behavior of Puppies," *Journal of Genetic Psychology* 77 (1950), 47.

38. Ibid., 50.

39. John P. Scott and John L. Fuller, *Genetics and the Social Behavior of the Dog* (Chicago: University of Chicago Press, 1965).

40. Ibid., 442.

41. Scott and Fuller's model "provides the basis for all texts published to date on this aspect of dog behavior." Miklósi, *Dog Behaviour, Evolution, and Cognition*, 303.

42. "Menzel [singular] (1937) also stressed the importance of the environment on the development of offspring. *He* [*sic*, emphasis added] presented a detailed description of the emerging attraction of dog pups to humans." Miklósi, *Dog Behaviour, Evolution, and Cognition*, 10.

43. The "breed warden" was responsible for inspecting the kennels of particular dog breeds and certifying each litter born there by recording the specific number of puppies in each litter and their genders and markings.

44. Blythe Hamer, *Dogs of War* (London: Carlton Publishing Group, 2001).

45. Binyamin Blum, "The Hounds of Empire: Forensic Dog Tracking in Britain and its Colonies, 1888–1953," *Law and History Review* 35, no. 3 (2017): 621–65.

46. Developed as a breed by the Germans in the late 1800s, Boxers are a cross between an older strain of German hunting dog used to chase down bears and wild boars and a smaller, mastiff-style dog from England. They are extremely loyal and loving but can be trained to be very vicious attack dogs. Originally bred to be hunters, they have exceptionally strong prey drives; when a Boxer catches its prey, its strong jaw locks like a trap and its powerful head and neck enable it to shake and tear at its victim until they are eviscerated. Because Boxers are so trainable and so fierce, the German and Austrian police began to use the breed following World War I to chase and catch fleeing suspects.

47. Resi Gerritsen and Ruud Haak, *The Malinois: The History and Development of the Breed in Schutzhund, Detection, and Police Work* (Edmonton: Brush Education, 2018), 121.

48. For more on how animal breeding acted as a metaphor for anxieties about nation, class, race, and gender in the nineteenth and twentieth centuries: Harriet Ritvo, *The Animal Estate: The English and Other Creatures in the Victorian Age* (Cambridge, MA: Harvard University Press, 2005).

49. Allport, Fay, and Hartshorne, "My Life in Germany Contest Papers," 114. Alverdes's formula was $A = f(k,v)$. "A" represents every biological action, and the formula illustrates the assumption that every living occurrence is the function of a constant factor "k," the inherited genetic material, as well as a variable factor "v," environmental influences.

50. Ibid., 114.

51. Ibid., 182.

52. There appear to be scant records of any direct links between Rudolphina and the Fortunate Fields Project, founded in Switzerland in 1924. The Fortunate Fields Project aimed to measure the development and analyze the psychological traits of a small population of German Shepherds. It was conducted only about five hundred miles away from Linz and took place at the same time as Rudolphina's work. The project was founded and funded by the American philanthropist Dorothy Harrison Eustis (1886–1946), who later established The Seeing Eye guide-dog school in Morristown, New Jersey, which Rudolphina visited in the 1950s after Eustis's death. See Elliot Humphrey and Lucien Warner, *Working Dogs* (Baltimore: John Hopkins Press, 1934; Wenatchee, Washington: Dogwise Publishing, 2005).

53. Robert Gibby and Michael Zickar, "A History of the Early Days of Personality Testing in American Industry: An Obsession with Adjustment," *History of Psychology* 11, no. 3 (August 2008), 164–84.

54. Rudolphina published a practical booklet on canine temperament testing in 1930, but all copies of the booklet have been lost according to Heinz Weidt and Dina Berlowitz, *Das Wesen des Hundes* (Augsburg: Naturbuch-Verl, 1998), 129. The original title of the Menzels' booklet was: *Wesenserprobung ihre theoretischen Grundlagen und ihre praktische Ausführung* (Augsburg: Association for German Shepherd Dogs, 1930).

55. Rudolph Menzel and Rudolphina Menzel, *Praktische Anleitung für die. Durchführung von Eignungsprüfungen bei den Nichtjagdhundrassen* [Practical Instructions for Carrying Out Aptitude Tests for Non-Hunting Dog Breeds] (Bern: Verlag Dr. Gustav Grunau, 1937). Translated by Susan Hecker Ray.

56. Rudolph Menzel and Rudolphina Menzel, *Ueber die Analyse hundlicher Charakteranlagen* [The Analysis of Canine Character Traits] (Berlin: Zeitschrift fur Hundesforschung, 1932), 194–95.

57. Canine temperament testing has since evolved into a variety of methods for recording and analyzing puppy responses to a variety of stimuli and provocations. Miklósi, *Dog Behaviour, Evolution, and Cognition*, 323–345.

58. Clarence Pfaffenberger's classic 1963 English-language publication *The New Knowledge of Dog Behavior* makes no mention of the Menzels' research despite lengthy discussions of canine temperament testing. Clarence Pfaffenberger, *The New Knowledge of Dog Behavior* (New York: Howell Book House, 1963). In John P. Scott and John L. Fuller's landmark 1965 publication *Genetics and the Social Behavior of the Dog*, no mention is made of the Menzels' work on canine temperament testing, though they do cite the Menzels' work on puppy development in their bibliography. Scott and Fuller, *Genetics and the Social Behavior of the Dog*. Steven Lindsay's 2001 publication on dog behavior and training—widely regarded as the most important publication since Scott and Fuller—supposedly includes a detailed review of the original research in the field but makes no mention of the Menzels' work. Steven Lindsay, *Applied Dog Behavior and Training: Etiology and Assessment of Behavior Problems* (Ames: Iowa State University Press, 2001), 2. A 2005 scholarly review of the literature on canine temperament-testing written in English also fails to include any mention of the Menzels' work: Amanda C. Jones and Samuel D. Gosling, "Temperament and Personality in Dogs (*Canis familiaris*): A Review and Evaluation of Past Research," *Applied Animal Behavior Science* 95, nos. 1–2 (2005): 1–53. German-language references to the Menzels' foundational work on canine temperament testing include: Bodingbauer, *Wesensanalyse für Junghunde*; and Eugen Seiferle and Emil Leonhardt, *Wesensgrundlagen Und Wesensprufung Des Hundes: Leitfaden Fur Wesensrichter* (Verlag, 1984).

59. Bodingbauer, *Wesensanalyse für Junghunde*.

60. Allport, Fay, and Hartshorne, "My Life in Germany Contest Papers," 110.

61. Ibid., 113.

62. Ibid., 111, 118.

63. Rudolph Menzel and Rudolphina Menzel, *Die Verwertung der Riechfähigkeit des Hundes im Dienste der Menschheit* (Berlin: Kameradschaft, Verlagsgesellschaft, 1930), 202.

64. Allport, Fay, and Hartshorne, "My Life in Germany Contest Papers," 115.

65. Ibid., 176; and Rudolph Wolf, *Der* Österreichische *Hundesport in Wort und Bild* (Wien: Redigiert, 1935), CZA 129/32.

66. Menzel, "Hounds of Hope," 8–9.

67. Wagner, *Jüdisches Leben in Linz*, 1072.

68. Ibid., 1074.

69. Allport, Fay, and Hartshorne, "My Life in Germany Contest Papers," 173–74.

70. Ibid., 173.

71. Ivan Pavlov, "Letter to Rudolphina Menzel," January 31, 1932 (in German), CZA A293/121.

72. Menzel, "Hounds of Hope," 8.

73. Allport, Fay, and Hartshorne, "My Life in Germany Contest Papers," 172.

74. Rudolphina claimed that the fact that many of them later became Nazis did not bother her, and she dismissed their political allegiances as primarily expedient, insisting that "dog sports" were free from politics. In fact, one of her closest friends from the Upper Austria Dog Club became the right-hand man to Theodor Habicht, one of the Nazi party's most enthusiastic leaders. Ibid., 106.

75. Allport, Fay, and Hartshorne, "My Life in Germany Contest Papers," 185.

76. Ibid., 107.

77. Rudolphina Menzel, "Raise Your Own Dog" (in Hebrew), *Habonim Dror Kibbutz Movement Youth Journal*, 1943, CZA A293/126.

78. Throughout the 1920s, both Menzels continued to give occasional lectures to the Zionist groups in Linz (e.g., Rudolph on the Babylonian exile and Rudolphina on the relationship of Jewish people to animals). Wagner, *Jüdisches Leben in Linz*, 1076.

79. Rudolphina Menzel, "General Survey on the History of Hebrew Dog-Breeding/Training in Eretz-Yisrael" (in Hebrew), August 8, 1953, HHA 33/29.

80. Rudolphina Menzel, "Letter to Arthur Ruppin" (in German), May 14, 1937, CZA A293/133.

81. Rudolphina Menzel "Use of Dogs in Army Service" (in German, English and Hebrew), CZA A293/4.

82. Rudolphina Menzel, "Roles for Dogs in Defense" (in Hebrew), CZA A293/99.

83. See Koriat in this volume.

84. Rudolphina's position on this issue had changed by 1937; photographs from her 1937 training course in Mikve Israel clearly show Jewish dog handlers being taught how to train attack dogs. Rudi Weissenstein, "Dog Training (Mrs. Menzel)," Mikve Israel, The National Library of Israel, system number 997000100670405171.

85. Phillip Ackerman-Lieberman and Rakefet Zalashik, *A Jew's Best Friend? The Image of the Dog throughout Jewish History* (Brighton, UK and Portland, OR: Sussex Academic Press, 2013).

86. Menzel, "Hounds of Hope," 8–9.

87. A brief account of the challenges involved in transporting three vigorous Boxer dogs by train and boat from Linz to Palestine at the time was provided by Hans Glaser:

> "...it was a catastrophe! At the border, when the Germans asked if we had cigarettes or money and checked our passports, I just gave the command and the dogs lunged at the door—the border officials didn't dare open it and didn't examine my papers at all...We then came to Trieste. The dogs were of course completely out of control and were very difficult to hold. I can only remember that we had to go through customs control there and I had to manage all three dogs. I wore mountaineering boots with studs on the bottom and I was dragged across the marble floor when the dogs saw another little dog until I found a pole to grab hold of." Wagner, *Jüdisches Leben in Linz*, 1099.

88. Rudolphina Menzel, "Letter," April 25, 1943, HHA, 290/34. Rudolphina noted her first trip to Palestine was at the explicit invitation of Yitzhak Ben Zvi and Yaakov Pat.

89. Allport, Fay, and Hartshorne, "My Life in Germany Contest Papers," 162. Rudolph was unable to join her on the 1934 trip to Palestine due to his professional commitments in Linz.

90. Allport, Fay, and Hartshorne, "My Life in Germany Contest Papers," 158–160.

91. Ibid., 160.

92. Ibid., 162.

93. Ibid., 133.

94. Ibid., 160.

95. Ibid., 125.

96. Ibid., 126.

97. Ibid., 162.

98. Rudolphina Menzel, "Travel Diary 1934," (German), HHA 290/34.

99. Rudolphina Menzel and Rudolph Menzel, *Dog Education and Training* (in Hebrew) (Palestine: Lanotter, 1939), 200.

100. Rudolphina Menzel, "Travel Diary 1934" (in German), HHA 290/34.

101. In 1934 the Mandate government passed a rabies ordinance to control outbreaks amongst both wild and domestic animals in Palestine; in relation to dogs, the ordinance allowed government officials free access to private premises in order to confiscate pet dogs suspected of being rabid, required all dogs to be quarantined after biting someone—on purpose or by accident—and mandated officials to kill dogs found to be rabid. It also permitted authorities to kill stray dogs at will. "Rabies Ordinance," *Official Palestine Gazette*, Schedule 1 (1934): 272.

102. *Hakalban* (August–September 1947): 39–40.

103. The Palestine police had not yet established their canine unit (which they did just a few weeks after Rudolphina returned to Austria in late 1934) when they imported three Doberman Pinschers from British-controlled Pretoria. See Blum, "The Hounds of Empire," 621–65.

104. As early as 1922, just two years after the establishment of the Mandate, British officers were hunting jackals with imported Fox Hounds, Harriers, and Airedales and keeping records of their outings and success, which from 1922 to 1926 amounted to 217 hunting days and 112 dead jackals. A. McNeill, "Hounds in Palestine," *The Palestine Post*, Friday, April 14, 1933; and "The Ramle Vale Hounds," *The Palestine Bulletin*, Friday, December 16, 1932.

105. While ubiquitous in early cynological literature, the term "pariah" has since been abandoned for designating street dogs because it was borrowed from the British colonialist term used to designate the most subordinated "outcast" social caste in India.

106. Rudolphina Menzel and Emil Hauck, "Letters," 1930s, CZA A293/133.

107. Allport, Fay, and Hartshorne, "My Life in Germany Contest Papers," 162.

108. All the leading cynologists of Europe were in attendance, including Professor Henseler of Munich, Professor Pirocci of Milan, Dr. Mery of Paris, and Konrad Most of Berlin. *Nature* (February 9, 1935): 228.

109. Allport, Fay, and Hartshorne, "My Life in Germany Contest Papers," 165.

110. Ibid., 166.

111. Ibid., 166.

112. Ibid., 166.

113. Ibid., 182.

114. Ibid., 167–68. Translation by Barbara Schmutzler and Susan Hecker Ray.

115. Wagner, *Jüdisches Leben in Linz*, 1099.

116. Sang-Hyun Lee, "Der deutsche Schäferhund und seine Besitzer. Zur Entwicklungs- und Bedeutungsgeschichte eines nationalen Symbols" (PhD diss., Tübingen, 1997). Hitler's German Shepherd named Blondi became a symbol of the Nazi animal rights movement.

117. Allport, Fay, and Hartshorne, "My Life in Germany Contest Papers," 186.

118. Ibid., 187.

119. The entire affair is described in great detail in Verena Wagner "*Der Menzelsche Boxerzwinger aus der Sicht einer Nachbarschaftsfehde* [The Menzel's Boxer Kennel through the Lens of a Neighborhood Feud]," in *Jüdisches Leben in Linz* (Linz: Wagner Verlag, 2008), 1076–87.

120. Allport, Fay, and Hartshorne, "My Life in Germany Contest Papers," 170.

121. Ibid., 169.

122. Rudolphina Menzel, "1937 Report on Work in the Land" (in German), HHA 290/34.

123. Rudolphina Menzel, "Report on Mikve Israel Dog Training Course 1937" (in German), CZA A293/4.

124. Since they opened their kennel in late 1934, Palestine police dogs had proven exceptionally effective in discovering perpetrators of "agrarian crimes such as crop burning, tree cutting, and animal maiming." Their success served as a model for British police all over the colonies. Blum, "The Hounds of Empire," 649.

125. Rudolphina Menzel, "1937 Report on Work in the Land" (in German), HHA 290/34. In a public notice that appeared in *The Palestine Post*, Monday, July 19, 1937: "This is to inform the public that on June 8, 1937 the formation of a Palestine Kennel Club was officially approved."

126. Rudolphina Menzel, "1937 Report on Work in the Land" (in German), HHA 290/34.

127. Rudolphina Menzel, "General Survey on the History of Hebrew Dog-Breeding/Training" (in Hebrew), August 8, 1953, HHA 33/29.

128. Ibid.

129. Allport, Fay, and Hartshorne, "My Life in Germany Contest Papers," 170.

130. Ibid., 70. The German pavilion was designed by Albert Speer—the chief architect of Nazi propaganda—and the Spanish pavilion featured Picasso's masterpiece *Guernica*, his testimony to the horrors of the Spanish Civil War.

131. Ibid., 170.

132. "Protokoll des Internationale Kynologen Kongresses Paris," CZA A293/3.

133. Menzel and Menzel, *Welpe und Umwelt*.

134. Allport, Fay, and Hartshorne, "My Life in Germany Contest Papers," 171.

135. Wagner, *Jüdisches Leben in Linz*, 1075.

136. Allport, Fay, and Hartshorne, "My Life in Germany Contest Papers," 190.

137. Allport, Fay, and Hartshorne, "My Life in Germany Contest Papers," 211.

138. Wagner, *Jüdisches Leben in Linz*, 1087.

139. Allport, Fay, and Hartshorne, "My Life in Germany Contest Papers," 195.

140. Ibid., 255.

141. As Rudolphina put it: "The thunderclouds of the coming world war already weighed heavily over us all." Ibid., 260.

142. Allport, Fay, and Hartshorne, "My Life in Germany Contest Papers," 256.

143. Allport, Fay, and Hartshorne, "My Life in Germany Contest Papers," 253.

144. Wagner, *Judisches Leben in Linz*, 1076.

145. Rudolphina Menzel, "The Dog in Service of Security and the Military" (Hebrew), no date, CZA, A293/80.

146. Avraham Tzur, "Hit'ahśadut ha-Yiśre'elit le-khalbanut," *Hakelev, Journal of the Israeli Kennel Club*, no. 9 (September 1973).

147. Interview with Ester Cohn, June 8, 2021.

148. Rudolphina Menzel, "Letter to Keren Kayemet" (Hebrew), October 21, 1939, CZA A293/15.

149. Allport, Fay, and Hartshorne, "My Life in Germany Contest Papers," front matter.

150. Rudolphina Menzel, "Appeal to the Jewish Agency to Organize Dogs for Security Services" (in German), CZA A293/83.

151. Menzel and Menzel, "Dog Education and Training."

152. Rudolphina and Rudolph Menzel, *Pariahunde*, translated by Bryna Comsky (Wittenberg: A. Ziemsen, 1960/2002), 53.

153. Ibid., 53.

154. M. Y. Ben-Gavriel, "Dogs Like Their School" and "Wonder of Taming," June 20, 1947, CZA A293/12.

155. Rudolphina Menzel, "On the Local Pariah Dog" (in Hebrew), undated, CZA A293/16.

156. Rudolphina Menzel, "Letter to Kibbutz Ramat Kovesh" (in Hebrew), October 18, 1939, CZA A293/13.

157. Rudolphina Menzel, "Annual Report" (in Hebrew), 1939, CZA A293/115.

158. Ibid.

159. Personal collection, Serena Fox.

160. Menzel and Menzel, *Pariahunde*, 5.

161. Rudolphina Menzel, "General Survey on the History of Hebrew Dog-Breeding/Training." *Palestine Kennel Gazette* (August/September 1946).

162. Menzel and Menzel, *Pariahunde*, 51.

163. Rudolphina's optimistic assessment of the pariah dogs' trainability was largely abandoned with time as the need for them became less acute. She learned that pedigree dogs born and raised in the country rather than imported to it were significantly better able to adapt to the climate. She also learned how to work with pedigree dogs at night when it was cooler. In subsequent years, as resources for importing more reliable European breeds increased, canine diseases became less prevalent, and knowledge about how best to care for pedigree dogs in a desert climate became more widespread, the need to train street dogs for service work was further obviated.

164. *Palestine Kennel Gazette* (November/December 1945).

165. *Palestine Kennel Gazette* (November/December 1945). "Our Stud Book contains as of October 31, 1945, 2083 registrations, grouped according to breed: 383 Alsatians, 119 Airdale Terriers, 1371 Boxers, 83 Brackenhounds, 15 Bullterriers, 132 'Canaan dogs,' 210 Dachshunds, 163 Doberman Pinschers, 44 Great Danes, 1 Pekinese, 43 German Pointers (both short and wire-haired), 23 Poodles, 7 Golden Retrievers, 23 Salukis, 14 Irish Setters, 88 Cocker Spaniels, 54 Scottish Terriers, 235 Wire-Haired Terriers, and 30 Tibetan Terriers."

166. Rudolphina Menzel, "General Survey on the History of Hebrew Dog-Breeding/Training."

167. Rudolph was not idle during this time, however. Among many other projects, he launched a study on tick-borne illnesses in tropical climates. Rudolph Reitler and Rudolph Menzel, "Dogs, Mice, and Ticks," *Transactions of the Royal Society of Tropical Medicine and Hygiene* 39, no. 6 (June 1946).

168. Interview with Ester Cohn, January 30, 2020.

169. Interview with Ester Cohn, January 30, 2020.

170. Allport, Fay, and Hartshorne, "My Life in Germany Contest Papers," front matter.

171. See Koriat in this volume.

172. Rudolphina was not unique in cultivating friendships with British Mandate officials and their wives. Many German-speaking Jews who arrived in Palestine in the fifth Aliya established friendly relations with the British. Tom Segev, *One Palestine Complete: Jews and Arabs under the British Mandate* (New York: Metropolitan Books, 2000), 488.

173. "Palestine Kennel Club Dog Show Program," CZA A293/18.

174. "Dogs Wanted for War Service" (English, German, Hebrew), CZA A293/64.

175. See Lehavi in this volume.

176. "Palestine Kennel Club Dog Show Program," HHA 14/52.

177. "Palestine Kennel Club Dog Show Program" and "A Dog Gymhkana: Program Aid Animal Hospital," CZA A293/3.

178. *Palestine Kennel Gazette* (April 1945).

179. Anonymous, "Letter from Haifa: Training Boxer Dogs," *The Palestine Review* 4, no. 1 (April 21, 1939), CZA A293/12.

180. Personal interview with Tslila Zagagi, November 2, 2020. Rudolphina Menzel, "Annual Report," 1945, CZA A293/4.

181. Dorothy Kahn Bar-Adon, "Boxers on the Battlefronts," *The Palestine Post*, Friday, July 13, 1945.

182. Leah Goldberg and Avigdor Rintso Lu'izada, *Yedidai me-reḥov Arnon* (Merḥavyah: Sifriyat po'alim), 1943.

183. Ernst Rettinger (1913–2009) was the proprietor of "All for the Dog." Born in Vienna, he and his wife Edith (1913–2008) arrived in Palestine shortly after the Menzels in 1939. "Czech Refugee Seeks Fellow Escapees," *The Jewish Floridian of Palm Beach County*, March 3, 1989, https://ufdc.ufl.edu/AA00014309/00128/12?search=rettinger.

184. *Palestine Kennel Gazette* (November/December 1945).

185. *Palestine Kennel Gazette* (May/June 1946).

186. *Palestine Kennel Gazette* (November/December 1945).

187. Kahn Bar-Ardon, "Boxers on the Battlefronts."

188. *Palestine Kennel Gazette* (June 1945). At the time, the Tel Aviv Dog Lovers Association had 230 members.

189. *Palestine Kennel Gazette* (March 1945).

190. Kibbutz Hazorea, on the western rim of the Jezreel Valley, was founded by German Jews in 1936.

191. *Palestine Kennel Gazette* (September/October 1944).

192. *Palestine Kennel Gazette* (August 1944). Rudolphina also described the best chemical compounds for extracting parasites and disinfecting kennels and listed substitutes that could be used during wartime shortages.

193. *Palestine Kennel Gazette* (December 1944).

194. Ibid.

195. *Palestine Kennel Gazette* (September/October 1944).

196. *Palestine Kennel Gazette* (June 1945). Rudolphina concluded that dogs have much better eyesight than is widely believed.

197. *Palestine Kennel Gazette* (May/June 1946).

198. *Palestine Kennel Gazette* (December 1944).

199. Rudolphina Menzel, *Thirty Years of Dog Breeding in Israel* (in Hebrew).

200. *Palestine Kennel Gazette* (September/October 1945).

201. *Palestine Kennel Gazette* (September 1944). Other lectures given by distinguished speakers at the monthly Dog Lovers Association meetings included: "Dogs in Archaeology," "Fauna of Palestine in Biblical Times and Now," "Breeding Problems in Palestine," and "Summer Diseases." Most of these talks were well-received, but only Rudolphina's appears to have been met with "stormy applause and cheers."

202. *Palestine Kennel Gazette* (February 1945).

203. *Palestine Kennel Gazette* (December 1944).

204. *Palestine Kennel Gazette* (April 1945).

205. *Palestine Kennel Gazette* (March 1945).

206. Ibid.

207. Ninety Boxers appeared in the April 1945 dog show, more than any other breed. *Palestine Kennel Gazette* (May 1945).

208. Ibid.

209. Purim, a sort of Jewish April Fool's Day, is an annual Jewish holiday that celebrates the survival of the Jews in Persia in the fifth century BCE.

210. Anonymous, "Palestine Kennel Club Invites Contacts," *The Palestine Post*, Friday, August 18, 1944, 8.

211. Kahn Bar-Ardon, "Boxers on the Battlefronts."

212. Benny Morris, *1948* (New Haven: Yale University Press, 2008), 87.

213. The Zionists were fighting with "black leopards" according to Egyptian newspaper reports during the war. Alexander and Ester Cohn, "Dogs in Tzahal" (in Hebrew, unpublished document), on file with the author.

214. Rudolphina Menzel, "General Survey on the History of Hebrew Dog-Breeding/Training."

215. Rudolphina Menzel, "Jobs of Dogs in the Army" and "About Distances in Training" (in Hebrew), CZA A293/99.

216. Ibid.

217. John Kistler, *Animals in the Military: From Hannibal's Elephants to the Dolphins of the U.S. Navy* (Santa Barbara, CA: ABC-CLIO, 2011).

218. A slip of paper with no date and no author described a patrol of thirty-six men and eight Boxers. The Boxers detected an ambush: "instead of losing 36 men, we lost 3 Boxers." CZA A293/4. The story also appears in the *Palestine Kennel Gazette* (August 1944).

219. British scent hounds had been uncovering caches of illegal ammunition since at least 1936, when Palestine police dogs discovered a large stock of explosives in an Arab village. In 1946, they uncovered stockpiles of ammunition belonging to Jewish resistance fighters as well. Blum, "The Hounds of Empire," 621–65.

220. Rudolphina Menzel "The Dog in Service of Security and the Military" (Hebrew), October 26, 1947, CZA A293/80.

221. Rudolphina Menzel, "General Survey on the History of Hebrew Dog-Breeding/Training."

222. Alexander and Ester Cohn, "Dogs in Tzahal."

223. Rudolphina Menzel, "The Dog as Partner to Haganah Activities" (in Hebrew), no date, HHA 33/28.

224. Lacking communication devices during the war, the Israeli forces relied on a clandestine network of sixty-eight dovecotes spread across the country and ran pigeoneer courses for soldiers. Rinat Harash, "Pigeons, Winged Warriors that Helped Israel to Victory 70 years ago," *Reuters*, accessed June 10, 2021, https://reut.rs/2ETmQSc.

225. Roski Zelig, "Note on Haganah budget for dog keeping in 1948 war: the budget was 25 liras per month" (in Hebrew), HHA 79.32.

226. See Koriat's description of the "Urabin Affair" in this volume.

227. Alexander Cohn and Ester Cohn, "Dogs in Tzahal."

228. Blum, "The Hounds of Empire," 621–65.

229. The history of the variegated relationship between Muslims and dogs, particularly in the Eastern Mediterranean, is beyond the scope of this discussion. For more, see: Cihangir Gündoğdu, "Dogs Feared and Dogs Loved: Human-Dog Relations in the Late Ottoman Empire," *Society & Animals* (2020): https://doi-org.ezp-prod1.hul.harvard.edu/10.1163/15685306-BJA10008. Also see Penny Johnson, *Companions in Conflict: Animals in Occupied Palestine* (New York: Melville House, 2019).

230. See Koriat in this volume.

231. *Hakelev*, no. 9 (September 1973). Interview with Ester Cohn, January 30, 2020. Cohn remembered being quite traumatized when, at the age of seventeen, Rudolphina demanded that she manually uncouple and clean the genitals of mating pariah dogs while surrounded by a group of men.

232. Ora Oren, "Psychological Treatment of Dogs," *He and She*, June 1, 1965, HHA 2379/80.

233. The first organized guide-dog school opened in 1916 in Oldenburg, Germany to train guide dogs for soldiers who lost their sight in World War I. Over five thousand guide dogs were trained at Oldenburg in its first twelve years, and soon, several other guide-dog schools were established in Western Europe. Riin Magnus, "Training Guide Dogs of the Blind with the 'Phantom Man' Method: Historic Background and Semiotic Footing," *Semiotica*, no. 198 (2014): 181–204, https://doi.org/10.1515/sem-2013-0107.

234. Rudolphina Menzel, *The New Outlook for the Blind* (New York: American Foundation for the Blind), 1964.

235. *Hakelev*, no. 9 (September 1973).

236. Poduschka-Aigner, "Menzel Rudolphine und Rudolph."

237. Rudolphina Menzel, *The New Outlook for the Blind*. In the 1950s, Israel contained a remarkably diverse population. There were almost 700,000 Jews in Palestine when the state was established in 1948; by 1951 that number doubled, as Jewish refugees streamed into the country from all over Europe, North Africa, and other regions of the Middle East. By the middle of the decade, an additional wave of Jewish immigration brought almost 200,000 more Jews to the country.

238. Rudolphina is listed as the Director of the Palestine Lighthouse in a 1954 to 1955 English-language publication of the organization, but her relationship with the organization ultimately soured for reasons that remain unclear. "What a Guide Dog Means to the Blind by Dr. Rudolphine Menzel, Director of Palestine Lighthouse Seeing Eye Foundation," CZA A293/12.

239. "Helen Keller's speech in Palestine advocating support for the blind," 1952, Hellen Keller Archive, American Foundation for the Blind, Series 2, Box 214, Folder 10.

240. Rudolphina's organization became eligible for American governmental funding after her niece Anita Boyko managed to register Friends of the Israel Foundation for Guide Dogs for the Blind as an American 501(c)(3) nonprofit organization in the United States (letter in personal collection of Serena Fox on file with the author). Research Project VRA-ISR-14-64.

241. Report on file with the author.

242. Ibid.

243. Rudolphina Menzel, G. Shapira, and E. Dreifuss, "A Proposed Test for Mobility-Training Readiness," *Journal of Visual Impairment & Blindness* 61, 2 (1967): 33–40, https://doi.org/10.1177/0145482X6706100201.

244. Milton D. Graham, "Wanted: A Readiness Test for Mobility," *The New Outlook for the Blind*, May 1965.

245. Rudolph and Rudolphina Menzel, *Pariahunde*, 51.

246. Rudolphina particularly wanted to generate data that explored Heini Hediger's distinction between the ontogenetic process of taming and the phylogenetic event of domestication. Rudolph and Rudolphina Menzel, *Pariahunde*, 43–45.

247. Brian Vesey-Fitzgerald, *The Book of the Dog* (Los Angeles: Borden Publishing Company, 1948), 977–78. The Menzels were not the first cynologists to express interest in street dogs. In *Pariahunde*, the Menzels quoted earlier writings by zoologists Theophile Studer, Otto Antonius, Emil Hauck, Max Hilzheimer, and Theodor Haltenorth.

248. Rudolph and Rudolphina Menzel, *Pariahunde*.

249. James Serpell's 1986 hypothesis was corroborated by a genetic study in 2009 that analyzed DNA from dogs around the world and found that all dogs belong to one genetic lineage, which suggests "pariah dogs" are not representatives of an intermediate evolutionary stage between wolves and dogs. Rather, they are more likely to be descendants of domesticated dogs that went feral and have adapted to their variable environments through natural selection. James Serpell, *In the Company of Animals: A Study of Human-Animal Relationships* (Oxford, New York: B. Blackwell, 1986). Ray and Lorna Coppinger offered a different hypothesis in 2001. They suggested that dogs did not descend directly from wolves, but rather, they are a separate species that domesticated themselves to exploit the ecological niche of Mesolithic garbage dumps. Ray Coppinger and Lorna Coppinger, *Dogs: A New Understanding of Canine Origin, Behavior, and Evolution* (Chicago: University of Chicago, 2001). For those thinking within a contemporary animal rights framework, capturing and domesticating freeranging dogs has come under intense criticism since it is based on a false premise about their evolutionary origins and interferes with their natural way of life.

250. Philosopher Donna Haraway describes the process of creating a breed standard thus: "A dog breeder educated in the traditional mentoring practices of the fancy will attempt through line breeding, with variable frequencies of outcrosses, to maximize the genetic or blood contribution of the truly 'great dogs' who are rare and special. The type is not a fixed thing, but a living, imaginative hope and memory." Donna Haraway, *When Species Meet* (Minneapolis: University of Minnesota Press, 2007), 148. Also see: Michael Worboys, Julie-Marie Strange, and Neil Pemberton, *The Invention of the Modern Dog: Breed and Blood in Victorian Britain* (Baltimore: Johns Hopkins University Press, 2018).

251. British hunters started breeding Greyhounds and Foxhounds in the eighteenth century, but it wasn't until the middle of the nineteenth century that denizens of the emerging English middle class started to selectively mate different types of dogs primarily as an amusing pastime. They discovered that dogs were a very plastic species, and it was not difficult to create novel varieties with different physical appearances. With a basic understanding of Mendelian genetics, a modern dog breeder could learn to make reliable predictions for determining the capacities of parents to pass on specific physical traits to their progeny. Experimenting with different combinations of dogs

to invent new physical forms soon became a popular amusement that quickly spread across Europe, even as the practice sometimes worked to create distance between the physical structure of a dog and its purported function. This new way of thinking about dogs was a departure from earlier practices when dogs were bred primarily for stamina or to possess particular functional capabilities. Modern dog breeding, by contrast, often produced dogs that were not necessarily structurally sound or behaviorally well-adapted, but who exemplified an aesthetic ideal or simply looked a certain way. The development of photography played a large role in the popularity of the new hobby of pedigree dog breeding because it meant images of model dog types could be circulated to promote conformity to new breed standards. The whole notion of standardization was itself a modern invention of the industrial age. As dog sports evolved, purebred dog breeding became competitive. The first dog show was held in England in 1859, but it wasn't until 1887 that a few exemplary breed standards were created, primarily for pointers, setters, and spaniels—effectively establishing individual breeds and breed standards. Breeders then vied with each other to produce individual dogs that were judged to best conform to these standards. "Winning" conferred prestige to the breeder as well as a financial reward as the sport became more popular.

252. Aaron Skabelund, *Empire of Dogs: Canines, Japan, and the Making of the Modern Imperial World* (Ithaca: Cornell University Press, 2011), 231.

253. Menzel, *Pariahunde*, 54.

254. Myrna Shiboleth, *The Canaan Dog* (Loveland, CO: Alpine, 1985), 12.

255. Ibid., 17.

256. Ibid., 17–36.

257. Ibid.

258. "History of the Canaan Dog," *The Canaan Dog Club of America*, accessed March 21, 2021, http://cdca.org/history.html.

259. Simon Bloom, "She Escaped the Nazis Who Commandeered Her to Train Dogs for Their Shock Troops," *American Jewish Ledger*, October 30, 1958.

260. Rudolphina Menzel and Rudolph Menzel, "Vorläufige Mitteilung Über Den Beziehungstype Hund-Katze," *Behaviour* 1, no. 3/4 (1948): 226–36, www.jstor.org/stable/4532687.

261. H. Strasser and W. Schumacher, "Breeding Dogs for Experimental Purposes: II: Assessment of 8-year Breeding Records for Two Beagle Strains," *Journal of Small Animal Practice*, 9 (1968): 603–12.

262. Rudolphina Menzel "Wert der Jugendveranlagungsprüfung," *Der Terrier* 2, no. 3 (1951), CZA A293/12. Rudolph Menzel and Rudolphina Menzel, "Europaische Hunderassen am Strand des Mittelbeeres und des Roten Meeres," *Schweizer Hundesport*, May 9, 1958, 202–204.

263. Bess Stephenson, "A Dog's Life is a Science to Israeli Woman Trainer," *The Courier-Journal*, July 17, 1955.

264. Personal correspondence with Myrna Shiboleth.

265. Rudolphina never saw her father again after she left Austria in 1938; he died in New York in 1944, and her stepmother died in 1947. Rudolph's parents died in Europe before the war (Erna Menzel, née Zucker, d. 1920; Ludwig Menzel, d.1936).

266. Interview with Serena Fox (Rudolphina's grandniece) on July 10, 2021.

267. Rudolphina's academic appointment was facilitated by two of her old friends from Europe, Henrik Mendelsohn and Hans Kreitler. Mendelsohn (1910–2002) was a Berlin-born alumnus of *Blau-Weiss* who founded the Zoology Department at Tel Aviv University and was one of the founders of the Society for the Protection of Nature in Israel. (Interview with Joseph Terkel, June 6, 2021). Kreitler (1916–1993) established the Psychology Department at Tel Aviv University and had been on the same boat with the Menzels when they all emigrated to Palestine from Austria in 1938. Interestingly, Yeshyahu Liebowitz, the renowned philosopher, taught physiological psychology in the same department at Tel Aviv University and was a colleague of Rudolphina's.

268. Rudolphina Menzel, January 14, 1968, CZA A293/112.

269. Interview with Joseph Terkel, June 6, 2021.

270. Rudolphina Menzel, "Letter to Dr. Ya'abetz at Tel Aviv University," February 28, 1966, CZA A293/112. The letter requested publication subvention for animal psychology textbooks and indicated commitment from Kibbutz Hameuchad publisher.

271. *Hakelev*, no. 9 (September 1973).

Two · Rudolphina's Early Years in Austria

1. Gordon Allport, Sidney Fay, and Edward Hartshorne, "My Life in Germany Contest Papers, 1940" (unpublished manuscript, 1940), printed volume, 3.

2. Frank Stern and Barbara Eichinger eds., *Wien und die jüdische Erfahrung, 1900–1938: Akkulturation, Antisemitismus, Zionismus* (Wien: Böhlau, 2009).

3. Allport, Fay, and Hartshorne, "My Life in Germany Contest Papers," 3.

4. Ibid., 3.

5. Ibid., 3.

6. Ibid., 6.

7. Robert Wistrich, *The Jews of Vienna* (Oxford, New York: Oxford University Press, 1989).

8. Allport, Fay, and Hartshorne, "My Life in Germany Contest Papers," 7.

9. Ibid., 37.

10. Ibid., 18.

11. *Österreichische Blau-Weiß für Jüdisches Jugendwandern* was established in 1913, and after merely five years, it boasted twenty-two local groups with 1,200 members. Inspired by the comparable German-Jewish *Wandervogel* groups, it sought to instil in its members a somewhat abstract notion of Jewish humanity, rather than a more specific type of Jewish consciousness.

12. Victoria Kumar, *Land und Verheißung: Ort der Zuflucht, Jüdische Emigration und nationalsozialistische Vertreibung aus Österreich nach Palästina 1920 bis 1945* (Innsbruck: Studieverlag, 2016), 83.

13. Kumar, *Land und Verheißung*, 83.

14. Allport, Fay, and Hartshorne, "My Life in Germany Contest Papers," 65.

15. Harriet Pass Freidenreich, *Gender and Identity: Jewish University Women in Vienna* (Bloomington: Indiana University Press, 2002), 300.

16. Alison Rose, *Jewish Women in Fin de Siècle Vienna* (Austin: University of Texas Press, 2008), 94.

Three · Rudolphina Menzel's Invention of Modern Dog Culture in Israel

1. Ephraim Benhar, "My Years with Dr. Menzel," *Hakelev*, no. 9 (September 1973).

2. *Hakalban*, no. 25 (June 1946).

3. Rachel Koriat, "'פנו דרך לחלוציות החדשה לכיבוש הכלב עבור בניין ארצנו: 'בניית שדה הפעולה' של גידול כלבים מקצועי בארץ-ישראל 1948–1934" [Pave the Way for the New Pioneering of the Occupation of the Dog for the Building of Our Country: The Construction of the Field of Action of Professional Dog Breeding in the Land of Israel (1948–1934)], The Department of Culture, Tel Aviv University, 2004. Master's thesis conducted in the Unit of Culture Research at Tel Aviv University under the supervision of Dr. Rakefet Sela-Sheffy.

4. Rudolphina Menzel, "Annual Report" (in Hebrew), 1939, CZA A293/27.

5. Ibid.

6. Rudolphina Menzel, "Annual Report" (in Hebrew), 1942, CZA A293/115.

7. Personal interview with David Efrat, May 2010.

8. Personal interview with David Efrat, May 2010.

9. Rudolphina Menzel, "Personal letter," CZA A293/126.

10. "Regulations of the Dog Lovers and Trainers Association in Eretz Israel" (in Hebrew), CZA A293/86.

11. Since dog shows held in 1940s Palestine were held in conjunction with officers from the British mandatory regime, they also provided a secondary benefit: they tightened the relationship between Jewish dog handlers and these officers.

12. Rudolphina Menzel, *Journal of the Dog Lovers and Trainers Association* (in Hebrew), CZA A293/18.

13. Rudolphina Menzel, "Annual Report" (in Hebrew), 1939, CZA A293/27.

14. *Hakalban*, no. 25 (June 1946).

15. Captain W. Foden, one of the British Mandate officers who worked with Rudolphina in Palestine, acted as the go-between with the Kennel Club of London and facilitated this recognition. CZA A293/27.

Notes to Chapter Three

16. Rudolphina Menzel, "Annual Report" (Hebrew), 1939, CZA A293/27.

17. Ibid.

18. Rudolphina Menzel, "Personal letter" (in Hebrew), CZA A293/16.

19. Rudolphina Menzel, "Annual Report" (in Hebrew), 1939, CZA A293/27.

20. Rudolphina Menzel, "Annual Report" (in Hebrew), 1942, CZA A293/115.

21. Notification of the event was published in *Hasade* (Volume B, 1922) and in a circular letter from the Hebrew organization in the State of Israel (Cycle B 23.X.1922N.111). The statement read as follows:

> On 8–9 January 1922 Haifa hosted the assembly of veterinarian physicians in the state of Israel, where the following main decision were made:
>
> 1. An association of the veterinary physicians in the state of Israel has been founded.
>
> 2. Since the veterinary practice is *organically associated with the general medicinal practice*, and since it is also preferable to avoid the unnecessary labor and expenses associated with the receipt of a special official approval for the association's existence—the association addresses the Hebrew physicians' association in the state of Israel with a request to annex the association as a "veterinary section" of the medical organization in the state of Israel [emphasis in original].
>
> 3. All of the veterinary physicians who are members of the association are obliged to be annexed to the medical organization in the state of Israel.
>
> 4. The association is called *The Association of Hebrew Veterinary Physicians in the State of Israel* [emphasis in original].

22. Dr. Rodolfson, "Letter," 1939, CZA A293/14.

23. Rudolph Menzel, "Letter," April 20, 1939, CZA A293/14.

24. Rudolphina Menzel, "Annual Report" (in Hebrew), 1939, CZA A293/115.

25. Rudolphina Menzel, "Letter," January 10, 1939, CZA A293/115.

26. *Hakalban*, no. 37–38 (June/July 1947).

27. Ibid.

28. Ibid.

29. Rudolphina Menzel, "Letter," November 16, 1945, CZA S53/1525.

30. Negotiations regarding the employment of Rudolph Menzel were mainly brokered by Arthur Ruppin.

31. Rudolphina Menzel, "Letter to Eliezer Kaplan," no date, CZA S53/1525.

32. Ibid.

33. Eliezer Kaplan, "Letter to Rudolphina Menzel," November 11, 1942, CZA S53/1525.

34. Rudolphina Menzel, "The English Attitude towards Dog Handling in Palestine," February 14, 1949, CZA 293/8 and HHA 34/289.

35. Shlomo Urabin, "The Dog in Security" (Hebrew), *Davar*, December 19, 1947.

36. The dubious term "dissident" was given to those who dissented from the organized, measured, and responsible resistance of the Yishuv against British rule in favor of extremist, violent, and irresponsible conflict. In other words, Urabin was similar to the "dissidents": a traitor who worked against the interests of the crowd he once belonged to. Rudolphina Menzel, "Letter," December 12, 1947, CZA A293/120.

37. Rudolphina Menzel, "Letter," August 5, 1947, CZA A293/120.

38. Ibid.

Four · Canine Zionism: Rudolphina Menzel and Working Dogs in Mandate Palestine

1. Rudolphina Menzel, *Dog Education and Training* (Palestine: Lanotter, 1939), 11–12, HHA 33/28: "The Jewish people were distant from nature during their long exile, during which they were primarily city dwellers and occupied in urban occupations. The dog was particularly [translation of the term דחוימב in Hebrew] foreign to us. He was the companion and assistant of the estate owner, the forester, and other gentiles from whose lifestyle Jews distanced themselves. This is how the canine was not given a place in our settlement project."

2. See Ora Oren, "Psychological Treatment of Dogs," *He and She*, June 1, 1965, HHA 2379/80.

3. Gideon, "Our Dogs Assist in Settling the Land," HHA 34/289.

4. Oren, "Psychological Treatment of Dogs."

5. Rudolphina Menzel, "Memorandum concerning the possibility of mobilizing the dogs of the Yishuv for the war-effort," no date, CZA S53/1525.

6. Rudolphina Menzel, "Memorandum," 1942, CZA S53/1525.

7. Rudolphina Menzel, "Secret Addendum, Memorandum," 1942, CZA S53/1525/31.

8. Rudolphina Menzel, "The English Attitude towards Dog Handling in Palestine," no date, HHA 34/289.

9. Dan Yahav, *Animals in the Security Services and the Haganah* (in Hebrew) (Israel: Cherikover Press, 2006), 85.

10. *Palestine Kennel Gazette*, no. 9 (February 1945): 1.

11. *Palestine Kennel Gazette*, no. 3 (August 1944): 3.

12. Ibid.

13. Rudolphina later reflected that she had hoped to train these dogs in novel ways to detect paratroopers and safe paths through minefields but had only succeeded in training them to detect mines. Rudolphina Menzel, *History of Dog Handling in Israel*, 6. HHA 33/29.

14. *Palestine Kennel Gazette*, no. 9 (February 1945): 1.

15. Rudolphina Menzel, "Memorandum" 1942, CZA S53/1525.

16. Rudolphina Menzel, "Report on the Arrest of a Murder Suspect," 1939, CZA A293/67.

17. Ibid.

18. Dare Wilson, *With 6th Airborne Division in Palestine, 1945–1948* (Barnsley, UK: Pen & Sword, 2008), 78–81.

19. Yahav, *Animals in the Security Services and the Haganah*, 86.

20. Kurt Wolf, "The Castration," HHA 34/289, 115–117.

21. Ibid.

22. Ibid.

23. Ibid. British authorities were able to find other eyewitnesses that had seen the culprits on their way to and from the operation. But when brought to a lineup, the victim allegedly refused to identify the culprits after one of the suspects whispered in his ear: "Haven't you had enough?"

24. AG v. Aref Ben Ahmad Esh Sharida, Crime No. 104/45 (Distict Court of Haifa, Sitting in Nazareth, July 4, 1945).

Five · Rudolphina Menzel's Contributions to the British War Effort

1. In 1939, the Germans laid 108,100 mines; in 1940, 102,100; in 1941, 220,900; in 1942, 1,063,600; in 1943, 3,414,000; and in 1944, they laid a record number of 8,535,500 mines. Mike Croll, *The History of Landmines* (Barnsley, U.K.: Leo Cooper, 1998), 41.

2. Michael Lemish, *War Dog: Canines in Combat* (Washington: Brassey's, 1996), 94.

3. Assistant Director of Veterinary and Remount Services "From the war diary of the headquarters in Jerusalem," June 2, 1942, The British National Archives (TNA), WO169/4344.

4. She later boasted that her methods were extremely efficient and enabled the detection of many mines. Rudolphina Menzel, "The Dog in the Security Services" (in Hebrew), no date, HHA, 33/28.

5. After the war, the British kept their promise. When the Palestine police used dogs to root out the Haganah's secret weapon stashes in the years leading up to the 1948 war, they used dogs that came from outside Palestine and were not trained according to Rudolphina's unique methods. Rudolphina Menzel, "Relationships with the British concerning dogs in Eretz Israel" (Hebrew), no date, HHA 33/28.

6. Rachel Koriat, "Pave the Way for the New Pioneering of the Occupation of the Dog for the Building of Our Country: The Construction of the Field of Action of Professional Dog Breeding in the Land of Israel (1948–1934)," The Department of Culture, Tel Aviv University, December 2004, 34–36. This describes the relationship between Dr. Menzel and Lieutenant Colonel E. Roger Newman. Also mentioned in two letters to Dr. Menzel from April and May of 1943, HHA 33/29.

7. Rudolphina Menzel, "Annual Report" (in Hebrew), 1943, HHA 33/29.

8. Anonymous, "War diary of Military Police dog training school," February 1, 1943, TNA WO169/13320.

9. No. 3 Remount Squadron, R.A.V.C. "War diary entries," March 27, 1943, TNA WO169/13430.

10. No. 3 Remount Squadron, R.A.V.C. "War diary entries," April 28, 1943, TNA WO169/13430.

11. Assistant Director of Veterinary and Remount Services, "Veterinary Report," December 7, 1943, TNA WO169/9039.

12. Ibid.

13. Ibid.

14. Ibid.

15. Rudolphina Menzel, "Relationships with the British concerning dogs in Eretz Israel" (in Hebrew), no date, HHA 33/28.

16. Dorothy Kahn Bar-Adon, "Boxers on the Battlefronts," *Palestine Post*, July 13, 1945. Edward V. Roberts, "Canine Mine Detectors," *Palestine Post*, October 27, 1944. Palestine Post Staff, "Mine-Dogs Trained by Jewish Experts," *Palestine Post*, September 9, 1946.

17. Rudolphina Menzel, "Annual Report 1947" (in Hebrew), HHA 33/28.

18. Ibid.

19. *Hakalban* (October/November 1947), HHA 291/34.

20. Yaakov Dori, "Letter to Dr. Menzel," February 19, 1948, HHA 33/28.

21. Zalman Cohen, "Letter to Rudolphina Menzel," June 7, 1948, CZA A293/4.

22. Ibid.

23. Rudolphina Menzel, "Letter to Moshe Dayan," March 16, 1948, CZA A293/29.

24. Avraham Zirlin, "Dr. Menzel's Way," *Hakelev*, no. 9, (September 1973).

25. Y. Mardi, "Letter to Rudolphina Menzel," May 7, 1951, Department of the British Commonwealth from the Foreign Secretary, CZA A293/80.

26. "From Professor Rudolphina to the 'Oketz' Unit-Dogs at the Top (Hebrew)," IDF Archives (website), retrieved February 17, 2022, https://archives.mod.gov.il/Exhib/animals/Pages/default.aspx.

27. Ibid.

Six · Personal Recollections of Rudolphina Menzel and Her Canaan Dog Breed

1. Rudolph and Rudolphina Menzel, *Pariahunde*, translated by Bryna Comsky (Wittenberg: A. Ziemsen, 1960/2002).

2. Private letter in personal collection of the author.

3. Rudolph and Rudolphina Menzel, *Pariahunde*, 1.

4. Private letter in personal collection of the author.

Seven · Rudolphina Menzel in Israeli Culture and Historiography

1. Rachel Koriat, "Pave the Way for the New Pioneering of the Occupation of the Dog for the Building of Our Country: The Construction of the Field of Action of Professional Dog Breeding in the Land of Israel (1948–1934)," Department of Culture, Tel Aviv University, December 2004. Lea Lehavi, "Dr. Rudolphina Menzel and the Military Kennel in the Jewish Community in Eretz Israel 1932–1948," Department of Jewish History, Tel Aviv University, February 2020. Opher Aderet, "Dogs of War," *Haaretz*, March 25, 2010.

2. Anonymous, "Center for the Blind will be Established in Israel," *Haaretz*, August 19, 1953.

3. Anonymous, "Evidence of Identification with the Help of Dogs—in the Bulkin Trial," *Herut*, November 15, 1965.

4. Rudolphina Menzel, "Friendship Means Life Together," *Maariv*, October 3, 1969.

5. David Tidhar, *Encyclopedia of the Founders and Builders of Israel* (in Hebrew), vol. 4, 1950, 4, 511.

6. *Hakelev*, no. 9 (September 1973).

7. Ester Cohn, "A Legend by the Name Menzel," Israeli Kennel Club (originally published 1987). During the retrieval of the article in July 2021, it emerged that this article was deleted from the website of the Israeli Kennel Club (it appeared on the site as of March 2020).

8. Alexander Cohn and Ester Cohn, "Dolphi and Rudy: The Menzels," *Animals and Society*, 30 (2006): 59–58.

9. "From Professor Rudolphina to the "Oketz" Unit-Dogs at the Top (Hebrew)," IDF Archives (website), retrieved February 17, 2022, https://archives.mod.gov.il/Exhib/animals/Pages/default.aspx.

10. "Rudolphina's Dogs (Hebrew)," Central Zionist Archives (website), Retrieved February 17, 2022, www.zionistarchives.org.il/en/AttheCZA/AdditionalArticles/Pages/Menzel.aspx.

11. "Israeli Center for Guide Dogs," Retrieved February 16, 2022, https://israelguidedog.org.il/en/history/.

12. "Rudolphina Menzel," *Wikipedia*, Retrieved on February 16, 2022, https://en.wikipedia.org/wiki/Rudolphina_Menzel.

13. Joan Wallach Scott, "Gender: A Useful Category for the Analysis of History," in *Ways of Feminist Thinking: A Reading*, eds. Nitza Yanai, Tamar Elor, Orly Lubin, Hanna Naveh, and Tami Amiel-Hauser (Raanana: The Open University, 2007), 264.

14. Billy Melman "Margins and the Center: A History of Women and the History of Gender in Israel," *Journal of the Israeli Historical Society* (2009): 245.

15. Aviva Halamish "The Biographical Age in Israeli Historiography," *Cathedra: On the History of the Land of Israel and its Settlement* 150, (2013): 241.

16. Ironically, dogs designated as "mixed Canaanite" are now amongst the most popular pets in Israel—the distinction is important. A "mixed Canaanite" is commonly understood to be a mongrel, not a purebred dog. Contemporary Israeli dog lovers, most often secular Israelis, readily embrace a "mixed" cultural symbol rather than a "purebred" one, perhaps reflecting a cultural rejection of the notion of a "pure" breed. This would make sense given Jewish experiences with the racist discourse of the Nazi regime. Moreover, there is a growing consciousness amongst secular Israelis about preventing cruelty to animals and about animal welfare in general; adopting mixed-breed dogs is a way for these Israelis to enact the virtues inherent in this discourse.

17. Shoham Smith, *True Legends: Fifty Women to Grow Up with in Israel*, ed. Yael Molchadsky (Dvir: Ḥevel Modi'in, 2019).

18. Rudolph Menzel and Rudolphina Menzel, *About Dogs, Cats and Other Friends* (Hebrew), (Tel Aviv: United Kibbutz Publishing, 1968), 6–7.

Select Bibliography of Rudolphina Menzel's Publications

All co-authored by Rudolph Menzel unless a different co-author is noted.

1913
Ernst Zerner and Rudolfine Waltuch. "Ein Beitrag zur Kenntnis der Pentosurie vom chemischen Standpunkt." *Monatshefte für Chemie* 34 (1913): 1639–52.

1914
Ernst Zerner and Rudolphina Waltuch."Zur Kenntnis der Pentosurie." *Monatshefte für Chemie* 35 (1914): 1025–36.

1920
"Dressur oder Erziehung." *Unsere Hunde* 6, no. 9 (1920).

1924
"Über die Camus'sche Einteilung des Boxergebisses." *Unsere Hunde* 10 (1924).
"Vom Meutenleben," I–III. *Unsere Hunde* 10, nos. 14, 16 and 23 (1924).

1925
"Über die Abrichtung des Boxers." *Unsere Hunde* 11, no. 4 (1925).
"Grundsätzliches zur Frage der Spürfähigkeit und Spurenreinheit." *Der Hund*, nos. 11 and 12 (1925).

1926
"Ketzergedanken zur Abrichtungsfrage." *Unsere Hunde* 12, no. 5 (1926).

1927
"Vorläufige Mitteilung über die Frage der Witterungsübereinstimmung." *Zeitung des Vereins für Deutsche Schäferhunde*, no. 14 (1927).

1928

"Grundsätzliches zur Frage der Spürfähigkeit und Spurenreinheit." *Der Hund*, nos. 11 and 12 (1928).
"Über die Maskierung des Hundecharakters durch die Normaldressur." *Unsere Hunde* 14, no. 5 (1928).
"Das Problem der Verwendung des Hundes als Feinriecher." *Der Deutsche Polizeihund* 28, no. 17 (1928).
"Gibt es eine Vererbung entwickelter Eigenschaften?" *Zeitung des Vereins für Deutsche Schäferhunde*, no. 3 (1928).

1929

"Beobachtungen über das Abstraktionsvermögen des Hundes." *Archiv für die gesamte Psychologie* 71, nos. 3 and 4 (1929).
"Die Bedeutung der Gesetze über Schwellenwert und Reizsummation bei der Spürarbeit des Hundes." *Journal für Psychologie und Neurologie* 38, nos. 3 and 4 (1929).
"Eigengeruch des Hundes als Fehlerquelle bei Witterungsreinheitsarbeiten." *Zeitung des Vereins für Deutsche Schäferhunde* (1929).
"Wesenserprobung, ihre theoretischen Grundlagen und ihre praktische Ausführung." *Zeitung des Vereins für Deutsche Schäferhunde* (1929).
"Zur Schutzhundfrage." *Der Hund*, nos. 14–16 (1929).
"Erziehung und Ausbildung des Hundes zum Begleit- und Schutzhund des Liebhabers." *Zeitung des Vereins für Deutsche Schäferhunde* 28 (1929).

1930

Die Verwertung der Riechfähigkeit des Hundes im Dienste der Menschheit. Wissenschaftliche Grundvoraussetzungen und praktische Anleitung zur Abrichtung u. Führung von Spürhunden im Ausforschungsdienst. Berlin: Kameradschaft, 1930.
Schwalbensommer. Vienna: Deutscher Verlag für Jugend und Volk, 1930.
"Wesenserprobung, ihre theoretischen Grundlagen und ihre praktische Ausführung." *Zeitung des Vereins für Deutsche Schäferhunde* (1930).
"Über die Größe des Boxers." *Sportblatt* (amtliches Organ des ZBB) 8, no. 4 (1930).
"Schönheit und Leistung." *Boxerblätter* 26, no. 11 (1930).

1931

"Über die Analyse hundlicher Charakteranlagen." *Zeitschrift der Gesellschaft für Hundeforschung* (1931).
"Der wehrhafte Hund." *Zeitung des Vereins für Deutsche Schäferhunde* (1931).
"Spezialisierung der Diensthundeabrichtung." *Zeitung des Vereins für Deutsche Schäferhunde* 30 (1931).
"Noch einmal zur Frage des Schutzanzuges." *Der Deutsche Polizeihund* 31, no. 2 (1931).
"Laienerfahrungen über die Stuttgarter Hundeseuche." *Berliner Tierärztliche Wochenschrift* 47, no. 27 (1931).
"Leistungszucht." *Der Hund*, no. 22 (1931).
"Erkenntnisse aus dem Gebiete der Boxerleistungszucht." *Boxerblätter* 27, no. 7 (1931).

1932

"Tierspsychologische Voraussetzungen für die Nasenarbeit des Hundes." *Zeitschrift für Hundefreunde* 2 (1932).
"Über die Analyse hundlicher Charakteranlagen." *Zeitschrift für Hundefreunde* 2 (1932).
"Was ist wichtiger, Leistungs-oder Anlageprüfung?" *Der Hund*, no. 2 (1932).
"Die Jugendveranlagungsprüfung." *Der Hund*, nos. 7 and 9 (1932).
"Vom Werdegang des Polizeifährtenhundes." *Der Hund*, no. 11 (1932).
"Leistungsprüfung." *Der Hund*, no. 12 (1932).
"Die Abrichtung des Hundes." *Der Hund*, no. 13 (1932): 15–18.
"Erziehung und Abrichtung." *Der Hund*, no. 22 (1932).
"Wesensgrundlagen." *Zeitschrift für Hundeforschung* 2, vols. 3 and 4 (1932).
"Praktische Anleitung für die Durchführung von Eignungsprüfungen bei den Nichtjagdhunderassen." *Zeitschrift für Hundeforschung*, nos. 3 and 4 (1932).
Kurze Bemerkungen zu dem Artikel "Der Boxer, seine Stellung als Diensthund und Dressur." *Boxerblätter* 28, no. 1 (1932).
"Die Kampftechnik des Boxers." *Boxerblätter* 28, no. 1 (1932).
"Die Reizverknüpfung, das 'Werkzeug' des Abrichters." *Der Hund*, no. 4 (1932).

1933

"Blut und Umwelt." *Der Hund*, no.7 (1933).
"Der Hund in der Stube." *Der Hund*, no. 24 (1933).
"Grundlehren der heutigen Abrichtetechnik." *Der Hund*, no. 6 (1933).
"Die Arbeit mit Witterungskonserven." *Zeitschrift für Hundeforschung* 3, nos. 1 and 2 (1933).
"Grundlehren der heutigen Abrichtetechnik." *Der Hund*, no. 6 (1933).

1935

Vererbung geistiger Eigenschaften beim Hund. Protokoll des Internationalen Kynologen-Kongresses Frankfurt/Main.

1936

"Parallelismen und Unterschiede im gesetzmäßigen Verhalten wilder und domestizierter Tiere." *Archiv für Psychologie* 97 (1936): 420–34.

1937

"Welpe und Umwelt." *Zeitschrift für Hundeforschung*, New Series, vol. 3 (1937).
Praktische Anleitung für die Durchführung von Eignungsprüfungen bei den Nichtjagdhunderassen. Berne: Buchdruckerei Gustav Grunau, 1937.
Hauptreferat über Tierpsychologie. Protokoll des Intern. Kynologen Kongresses Paris.

1939

Dog Education and Training (in Hebrew). Tel Aviv: Lanotter, 1939.
"Einiges über den Paria." *Schweizer Hundesport* 55, no. 6 (1939).
Leitfaden für Hundehaltung und Abrichtung, hrsg in und für Israel.

1948

"Observations on the Pariah Dog." In *The Book of the Dog*, edited by Brian Vesey-Fitzgerald, 968–90. London-Brussels: Nicholson & Wattson, 1948.

"Vorläufige Mitteilung über den Beziehungstyp Hund-Katze." *Behaviour* 1, Part 3–4 (1948): 226–36.

1949

"Kynologie im Aufbau." *Unsere Hunde* 26, no. 4 (1949): 7–10.

1951

"Wert der Jugendveranlagungsprüfung." *Der Terrier*, nos. 2 and 3 (1951).

"Blut und Umwelt." *Unsere Hunde* 28, no. 2 (1951): 1–5.

"Canaan-dog und österreichischer kurzhaariger Pinscher." *Unsere Hunde* 28, nos. 7/8 (1951): 2–3.

1952

"Der Wesensstandard, eine notwendige Ergänzung zum Formstandard." *Schweizer Hundesport* (1952).

1953

"Einiges aus der Pflegewelt der Mutterhündin." *Behaviour* 4 (1953): 289–304.

"Kynlogisches aus Israel." *Unsere Hunde* 30, vol. 3 (1953): 4–6.

1954

"Angeborene und erworbene Schemata beim Haushund." *Unsere Hunde* 31, no. 6 (1954): 2–8; nos. 7/8 (1954): 2–7.

1957

"Vom Paria zum Canaan Dog." *Schweizer Hundesport*, no. 9.

1958

"Europaische Hunderassen am Strand des Mittelmeeres und des Roten Meeres," *Schweizer Hundesport*, May 9, 1958, 202–04.

1960

Pariahunde. Wittenberg: A. Ziemsen Verlag, 1960.

1963

"Unser Boxer, wie er war und wie er ist."

"Die erbbiologische Bedingtheit des Boxerwesens." *Der Boxer*. Vienna: Österr. Boxerklub, 1963.

1964

"Erfahrungen über Hundezucht und Hundehaltung im subtropischen Klima." *Unsere Hunde* 41, no. 3 (1964): 4–9.

1967

Rudolphina Menzel, G. Shapira, and E. Dreifuss. "A Proposed Test for Mobility-Training Readiness." *Journal of Visual Impairment & Blindness* 61, no. 2 (1967): 33–40.

1968

About Dogs, Cats, and Other Friends (in Hebrew). Tel Aviv: United Kibbutz Publishing, 1968.

Contributors

MONIKA BAÁR is professor by special appointment of Central European Studies in the Institute of History at Leiden University. She received a doctorate in modern history from the University of Oxford (2002) and subsequently held a two-year postdoctoral fellowship at the Max Planck Institute for the History of Science in Berlin and a two-year teaching fellowship at the University of Essex. Before joining the Institute for History at Leiden University in 2015, she was Rosalind Franklin Fellow at the University of Groningen between 2009 and 2015. Her current research interests revolve around the history of disability and the history of the human-animal relationship.

TAMMY BAR-JOSEPH is pursuing her MA in cultural studies at the Open University in Israel and serves as a member of the steering committee of the Human-Animal Relations Research Forum at Tel Aviv University. Her master's thesis focuses on Jewish children rescued by dogs in the Holocaust. She published "Nazis, Dogs, and Collective Memory: The Holocaust's Impact on Negative Attitudes toward Dogs in Jewish Society in Israel" in *Moreshet: Journal for the Study of the Holocaust and Antisemitism* (2019). With Hadas Marcus, she published "From Trauma to Trust: The Convoluted Relationship Between Jews and Dogs" in *Animals and Ourselves: Essays on Connections and Blurred Boundaries* (McFarland, 2020). In 2020, she received a Presidential Certificate of Excellence from the Open University of Israel, and in 2021, she received a Research Encouragement Scholarship from the Yad Vashem Scientific Committee. As an educational-therapeutic dog trainer, she works with high school students and gives support and guidance to families who raise dogs.

BINYAMIN BLUM joined the University of California in Hastings faculty in spring 2018. Prior to coming to Hastings, Professor Blum was on the Law Faculty of the Hebrew University in Jerusalem from 2012 to 2017. As a legal historian of the British Empire, Blum specializes in the relationship between law and colonialism during the nineteenth and twentieth centuries. Blum also writes on current issues of evidence and proof such as the exclusion of unlawfully obtained evidence, spousal privilege among same-sex partners, rape shield statutes, and character evidence. After receiving his BA and LLB (summa cum laude) from the Hebrew University of Jerusalem, Blum clerked for the Honorable Justice

Ayala Procaccia of the Israeli Supreme Court. He went on to earn a doctorate in law and an MA in history as a Presidential Fellow at Stanford University. In 2009, Blum was a fellow at the J. Willard Hurst Institute of Legal History at the University of Wisconsin in Madison. At Hebrew University, Professor Blum taught legal history, law, and colonialism and evidence. He co-chaired the Jerusalem Legal History Forum, which runs a biennial workshop, and the Jerusalem Crime Group, an interdisciplinary forum for law enforcement policy analysis. He is the co-founder of the British Colonial Legalities Collaborative Research Network in the Law and Society Association. In 2013 and 2014, he served as a visiting professor at Stanford Law School.

SUSAN MARTHA KAHN is the associate director of the Julis-Rabinowitz Program on Jewish and Israeli Law at Harvard Law School. She received her PhD in anthropology and master's degree in middle eastern studies from Harvard University. She has published in science studies, animal studies, and Jewish studies, and her book *Reproducing Jews: A Cultural Account of Assisted Conception in Israel* (Duke University Press, 2000) won a National Jewish Book Award as well as the Eileen Basker Prize, from the American Anthropological Association, for Outstanding Research in Gender and Health.

RACHEL KORIAT was raised by her parents, Pnina (née Volanov) and Shmuel Tzvieli-Toretzky on Kibbutz Tirat Zvi in the Beit She'an Valley. From 1962 to 1964, she served as an education officer in the Nahal Brigade of the Israel Defense Forces (IDF). After completing her military service, she left the kibbutz and began her long journey in education and teaching. She received a BA (with honors) in Hebrew language, Hebrew literature, and political science from the University of Haifa, and an MA (with honors) from Tel Aviv University. Her master's thesis on Dr. Rudolphina Menzel is the basis for her article in this book. She has held roles in teaching, education, management, and teacher training in various educational frameworks, and she is married to Moshe Koriat, with whom she shares four children and six grandchildren.

LEA LEHAVI received her BA in world history and Israel studies and her MA in Jewish history from Tel Aviv University. Her master's thesis, *Dr. Rudolphina Menzel and the Military Canine Formation in the "Yishuv" during Mandatory Palestine 1932–1948*, was supervised by Professor Motti Golani. She also served in the IDF's elite canine unit Oketz.

MYRNA SHIBOLETH has been breeding Canaan dogs since 1969 and has produced over fifty dogs that have claimed titles, including International Champion and World Winner. She is very proud that her dogs provide the basis for most of the Canaan dogs around the world. She is considered the world authority on the breed and has lectured on the Canaan dog in Israel, Finland, France, England,

Italy, Kazakhstan, and the United States. She became a judge of Collies in 1971 and is now an Fédération Cynologique Internationale (FCI) judge of all breeds. She has judged in the United States, Israel, Sweden, Norway, Finland, Luxembourg, Poland, Germany, Cyprus, Greece, Belgium, Croatia, Slovenia, Italy, Estonia, Latvia, Russia, France, Australia, and Denmark. She is a past member of the board of the Israel Kennel Club, and past president of the Israel Spitz Dog Club. She frequently writes for various dog publications in Israel and abroad, and was a regular contributor to *Collie Revue* in Germany. Among her publications are the breed guide *The Israel Canaan Dog* (first and second editions, Loveland, CO: Alpine Publications, 1985 and 1996; third edition, Hendelmade, 2012) and *Tails of Shaar Hagai: A Wild Life with Wildlife* (Durham, NH: Sephirot Press, 2008). She regularly lectures on a wide variety of canine subjects, including behavior, structure and movement, communication, service and therapy dogs, training methods, and breeding, both in Israel and abroad; and she has given the first advanced course in canine subjects for the judging aspirants of the Cyprus Kennel Club. Overall, she lives with and enjoys her dogs.

Index

Page numbers with an "n" indicate an endnote.

Aaronson, Sarah, 134
Adler, Victor, 7, 97
Allied Forces, 108, 112–13, 119, 132
Alverdes, Friedrich, 18–19
American Kennel Club (AKC), 78–79
Animals and Society, 132
Arab(s), 56, 67–68, 93, 102–3, 109, 111, 134; fellahin, 54; fighters, 69; horses, 112; hunting hounds, 39; poetry, 39; populations, 65; resistance to Zionist settlement, 32–33, 66; sabotage of, 112
Arab Revolt, 44, 115
Arbeiter-Zeitung, 6
Austrian Boxer Club, 27, 30
Austrian Cynology Association, 27, 30

Bedouin, 39, 77–79, 124–25, 128, 134
Beer-Hoffmann, Richard, 7
The Behavior of Animals in Zoos and Circuses (Hediger), 80
Benhar, Ephraim, 91
Ben-Zvi, Yitzhak, 34, 38
Berkowitz, Ursula, 126
Blau-Weiss (Zionist organization), 8–10, 63, 88–89, 155n267
B'nei Habitachon (Children of Security), 78, 126
B'nei Satan (Children of Satan), 30
Bodingbauer, Joseph, 11, 13, 17, 19, 25, 28
Bonem, Paul, 62–63
The Book of the Dog (Vesey-Fitzgerald), 75
Boxers, 14, 44, 50, 142n46, 151n218; breed/breeding of, 18, 31, 34, 53;

Linzer, 25; perseverance of, 106; puppies, 15; purebred, 17; trained to detect land mines, 113; trained to detect visitors (and intruders), 111; training of, 112
Boyko, Anita, 152n240
Boyko, Helene (Waltuch), 80
British Mandate, 63, 68, 105, 146n104, 149n172; colonial subjects of, 33; officials, 39, 51, 56, 105, 149n172, 156n15; police dogs, 68; policemen, 57; public healthcare, 38; Veterinary Services Department, 119
British Red Cross, 63
Buber, Martin, 8

Canaan dog(s), 69–81, 94, 123–35, 142n52, 152n233
Canaan Dog Club of America, 80
Canis familiaris, 53, 75
Central Zionist Archives, 132
Chayes, Zevi Hirsch, 29
Child and Marriage Guidance Agency, 35
Cohn, Alexander, 68
Cohn, Ester, 68, 132, 139n3, 152n230
Coppinger, Lorna, 153n249
Coppinger, Ray, 153n249
Corps of Royal Military Police, 115
Criminal Investigation Department (CID), 114

Davar, 106–7, 131
Dayan, Moshe, 65, 121–22

Index

Der Hund, 42
Deutscher Turnverein (the German Sports Club), 87
Die Welt, 4, 87
dissident, 107, 158n36
Dobermann, Karl Friedrich Louis, 76
Dog Education and Training (Menzel), 110
Dog Lovers and Trainers Association in Palestine (DLAT), 94–97; first year of operations, 100–105
Dog Lovers Associations, 56, 59, 63, 65, 79, 95, 96, 105, 149n201
Dog Ten Commandments (Menzel), 64
Dori, Yaakov, 121–22

Efrat, David, 93
Ein HaHoresh Kibbutz, 93
Eisner, Freia, 126
Encyclopedia of the Pioneers of the Yishuv and its Builders, 131
English Kennel Club, 127
Eshkol, Levi, 67
Eustis, Dorothy Harrison, 142n52

Federation Cynologique Internationale, 30, 125
Feuchtinger, Friedrich, 27–28
Foden, W., 156n15
Fortunate Fields Project, 142n52
Frank, Anne, 140n8
Freie Schule Association, 5, 11
Freud, Sigmund, 3, 14, 16
Friends of the Israel Foundation for Guide Dogs for the Blind, 152n240
Fuller, John L., 17, 143n58

German Shepherds, 29–30, 42, 46, 71, 76, 106, 146n116
Gies, Miep, 140n8
Glaser, Hans, 145n87
Glockel, Leopoldine, 5–8, 11, 35, 44, 140n6
Glockel, Otto, 6–7, 35
Goldberg, Leah, 59
Goldschmidt, Martin, 59, 95

Goring, Herman, 50
Great Danes, 57
Greyhounds, 153n251
Gross, Hans Gustav Adolph, 18
Guernica, 147n130

Haaretz, 131
Habicht, Theodor, 144n74
Habimah Theater, 40
Hadassah, 103
Haganah, 34, 38, 66–67, 91, 114–116, 121–122, 132, 159n5
Hakalban (The Dog Handler), 59–60, 62, 64, 101–2
Hakelev, 132
Haraway, Donna, 153n250
Hashomer (Jewish defense organization), 32
HaTikva Canaans, 78
Hauck, Emil, 17–18, 19, 28, 39, 75–76
Hediger, Heini, 80, 153n246
Herut, 131
Herzl, Theodor, 4, 8
Hexter, Maurice Beck, 37
Higgins, Connie, 126–27
Hilzheimer, Max, 29
Hippocrates, 16
Hitler, Adolf, 29, 35, 42, 48, 50, 88, 105
Holocaust, 68–69

Institute for the Orientation and Mobility of the Blind, 70–74, 80, 123
Institute Pasteur, Tel Aviv, 97
International Cynological Congress, 30, 41, 43, 47, 132
International Cynological Federation, 96–97
Irgun, 45, 114, 158n36
Israel Defense Force (IDF), 122, 132, 135
Israeli Center for Guide Dogs, 132–33
Israeli Kennel Club, 132

Jewish Agency, 34, 37–38, 40, 44–46, 51, 56, 102, 106–7, 111, 119
Jewish Gymnastics Club, 4, 88
Jewish National Fund (JNF), 80, 129

Index

Jewish Soldiers Welfare Committee, 63
Joint Distribution Committee, 72
Jüdische Wanderbund Blau-Weiss. See Blau-Weiss
The Jungle Book (Kipling), 12

*K*aftanjuden, 86–87
Kaplan, Eliezer, 102–104
Keller, Helen, 72
Kennel Club of London, 97, 156n15
Kipling, Rudyard, 12
Kiryat Haim, 67, 71, 80, 128
Kiryat Motzkin, 51–54, 57–58, 67, 69, 77, 92, 98, 107, 111, 120, 132
Kleinmunchen, 13–14, 25, 27–28, 31, 34, 41–42, 44, 47, 49, 52, 58
Kreitler, Hans, 155n267
Krikorian, K. S., 38

*L*aish me B'nei Habitachon, 128
Lassie Come Home, 63
Lehi resistance, 158n36
Leopoldstadt, 86
Lindsay, Steven, 143n58
Lindtberg, Leopold, 40
Loeb, Ernest, 30
Lorenz, Konrad, 28
Lueger, Karl, 87
"Lullaby for Miriam" (Beer-Hoffmann), 7

*M*aariv, 131
Maccabi Stadium, Tel Aviv, 63
Magen David Adom 107
Mahler, Gustav, 3
Mann, Thomas, 13
Marston, Mary Vesta, 17
Mazel Tov Canaan Dogs, 78
Menzel, Rudolph, 9–11, 90, 140n19
Menzel, Rudolphina, 58, 142n42; about death, 81; about parents, 3, 5; about stepsiblings, 80; assisting the Palestine Police, 114; contributions to the British war effort, 119–22; creation of the mobility readiness test, 74; early exposure to/passion for Zionism, 4, 9; early years in Austria, 85–90; introduced to socialism, 5; invention of modern dog culture in Israel, 91–108; in Israeli culture and historiography, 131–35; Jewish underground movements, 114–17; personal recollections of, 123–29; professional dog breeding and training, 92–93, 105, 113; public lectures delivered on Zionism, 6–7, 52, 114–17; studies at the University of Vienna, 8–9
Mesilot Kibbutz, 116
Mikve Israel, 38, 91, 144n84
mine-detecting, 54, 57, 66, 96, 115, 120; in North Africa, 119; training of dogs for, 32, 55, 93, 119, 121;
mobility readiness test, 74
Most, Konrad, 4, 12, 26, 28, 37, 141n27
My Friends from Arnon Street (Goldberg), 59

*N*euberger, Yitzhak, 62
Nuremberg Laws, 36

*O*ketz, 122, 135

*P*alestine Kennel Club, 45, 55–57, 59, 147n125
The Palestine Kennel Gazette, 64
Palestine Lighthouse, 72–73, 152n238
Palestine Post, 58, 121
Palestine Research Institute for Canine Psychology and Training, 52, 69
Parasitology Institute at Hebrew University, 97–98
"pariahs," 39, 75, 125, 127, 146n104
Pariahunde, 75, 125
Paris World's Fair, 47
Pat, Yaakov, 34, 38, 122, 132
Pavlov, Ivan, 14, 16, 28

*R*eitler, Rudolph, 98
Rettinger, Edith, 149n183
Rettinger, Ernst, 149n183
Roger, E. N. Newman, 120
Ruppin, Arthur, 32, 102

Index

Santrouschitz, Hermine, 140n8
Schwabacher, Joseph, 29
Scott, John Paul, 17
Seeing Eye guide-dog program, 80
Seiferle, Eugen, 28
Seitz, Karl, 7
Serpell, James, 153n249
Service Dogs Handlers Union, 107
Sha'ar Hanegev Kibbutz, 46
Shaar Hagai, 127
Sharett, Moshe, 38, 57, 105, 108, 112, 119
Sharida, Aref, 115–17
Skinner, B. F., 12
Social Democratic Party, 6–8, 10, 31, 35, 88
Society for the Preservation of the Japanese Dog, 76
SPCA Tel Aviv-Jaffa, 63
Speer, Albert, 147n130
Spring House breeding kennel, 93
Strauss, Peter, 34
Swiss Cynological Society, 44
Szenes, Hannah, 134

Technische Hochschule, 89
Temperament test, 21-25
Tirat Zvi Kibbutz, 91
True Legends: Fifty Women to Grow Up with in Israel, 135
Tur-Sinai, Naphtali Herz, 37, 80
Tzur, Avraham, 51, 132
Tzvieli-Toretzky, Shmuel, 91

United Nations Partition Plan, 67
United States Department of Health, Education, and Welfare, 73

University of Vienna, 8, 12, 89–90
Urabin, Shlomo, 105–6
Urabin affair, 105–8

Verein Zionistischer Hochschüler Theodor Herzl (Theodor Herzl Jewish Students Association), 90
Vesey-Fitzgerald, Brian, 75
Veterans Department, 71
Viktor, Adler, 140n9
Vitales, Rose, 37
von Donnhauf Kennel, 12
von Schonerer, Georg Ritter, 140n9
von Stephanitz, Max, 30, 76
von Tschammer-Osten, Hans, 41

Waltuch, Egon, 80
Waltuch, Reb Fischl, 86
Wandervogel, 89, 155n11
Warden, Geoffrey, 39
War of Independence, 92, 131
Welpe und Umwelt (Puppy and Environment), 47
Wittgenstein, Ludwig, 3
Wolf, Friedrich, 40
World Cynological Congress, 30, 43, 47

Yagur Kibbutz, 38
Yishuv, 31–34, 57, 92, 103–5, 119, 121, 131, 158n36; dog-training courses, 93; lacking communication devices, 93; leaders, 38, 65, 102, 158n36; security of, 107

Zerner, Ernst, 9
Zirlin, Abraham, 68, 122

The Tauber Institute Series for the Study of European Jewry

JEHUDA REINHARZ, General Editor
CHAERAN Y. FREEZE, Associate Editor
SYLVIA FUKS FRIED, Associate Editor
EUGENE R. SHEPPARD, Associate Editor

The Tauber Institute Series is dedicated to publishing compelling and innovative approaches to the study of modern European Jewish history, thought, culture, and society. The series features scholarly works related to the Enlightenment, modern Judaism and the struggle for emancipation, the rise of nationalism and the spread of antisemitism, the Holocaust and its aftermath, as well as the contemporary Jewish experience. The series is published under the auspices of the Tauber Institute for the Study of European Jewry—established by a gift to Brandeis University from Dr. Laszlo N. Tauber—and is supported, in part, by the Tauber Foundation and the Valya and Robert Shapiro Endowment.

For the complete list of books that are available in this series, please see www.brandeis.edu/tauber

SUSAN MARTHA KAHN
Canine Pioneer:
The Extraordinary Life of
Rudolphina Menzel

GILAD SHARVIT
Dynamic Repetition:
History and Messianism in
Modern Jewish Thought

YOSEF HAYIM YERUSHALMI
in Conversation with
SYLVIE ANNE GOLDBERG
Transmitting Jewish History

ARTHUR GREEN
Defender of the Faithful:
The Life and Thought of
Rabbi Levi Yitshak of Berdychiv

CHARLES DELLHEIM
Belonging and Betrayal:
How Jews Made the
Art World Modern

CEDRIC COHEN-SKALLI
Don Isaac Abravanel:
An Intellectual
Biography

www.ingramcontent.com/pod-product-compliance
Lightning Source LLC
Chambersburg PA
CBHW060355080526
44583CB00012B/326